The Dolciani Mathematical Expositions

NUMBER THIRTY-ONE

A Garden
of
Integrals

Frank E. Burk
California State University at Chico

Published and Distributed by
The Mathematical Association of America

DOLCIANI MATHEMATICAL EXPOSITIONS

Committee on Publications
James Daniel, *Chair*

Dolciani Mathematical Expositions Editorial Board
Underwood Dudley, *Editor*
Benjamin G. Klein
Virginia E. Knight
Mark A. Peterson
Robert L. Devaney
Tevian Dray
Jerrold W. Grossman
James S. Tanton

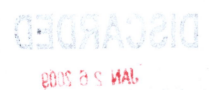

The DOLCIANI MATHEMATICAL EXPOSITIONS series of the Mathematical Association of America was established through a generous gift to the Association from Mary P. Dolciani, Professor of Mathematics at Hunter College of the City University of New York. In making the gift, Professor Dolciani, herself an exceptionally talented and successful expositor of mathematics, had the purpose of furthering the ideal of excellence in mathematical exposition.

The Association, for its part, was delighted to accept the gracious gesture initiating the revolving fund for this series from one who has served the Association with distinction, both as a member of the Committee on Publications and as a member of the Board of Governors. It was with genuine pleasure that the Board chose to name the series in her honor.

The books in the series are selected for their lucid expository style and stimulating mathematical content. Typically, they contain an ample supply of exercises, many with accompanying solutions. They are intended to be sufficiently elementary for the undergraduate and even the mathematically inclined high-school student to understand and enjoy, but also to be interesting and sometimes challenging to the more advanced mathematician.

MAA Service Center
P.O. Box 91112
Washington, DC 20090-1112
1-800-331-1MAA   FAX: 1-301-206-9789

To

Don Albers

*for all his years of
dedicated service to the MAA*

Foreword

From quadratures of lunes to quantum mechanics, the development of integration has a long and distinguished history. Chapter 1 begins our exploration by surveying some of the historical highlights—milestones in man's capacity to think rationally. Whether motivated by applied considerations (areas, heat flow, particles in suspension) or aesthetic results such as

$$\frac{1}{\pi} = \frac{\sqrt{8}}{9801} \sum_{n=0}^{\infty} \frac{(4n)!}{(n!)^4} \frac{1103 + 26390n}{396^{4n}},$$

the common thread has been and will continue to be *understanding*.

Mathematical discoveries are but markers in our quest to understand our place in this universe. In the profession of mathematics, we are all too frequently humbled, but we persevere for those rare moments of exhilaration that accompany understanding. That is the nature of mathematics—indeed, the nature of understanding.

I hope that you will personally experience the emotional peaks and valleys that are presented in the pages that follow. If we attain a measure of understanding and an appreciation of our mathematical ancestors and their accomplishments, our efforts will have been successful. We will have played a role in our inexorable journey to the stars.

— F.B.

Contents

CHAPTER 1

An Historical Overview

Every one of us is touched in some way or other by the problems of mathematical communication. Every one of us can make some contribution, great or small, within his own proper sphere of activity. And every contribution is needed if mathematics is to grow healthily and usefully and beautifully. — E. J. McShane

1.1 Rearrangements

Figures 1(a) through (d) demonstrate the general idea of rearranging a given area to form another shape. In the first example, we have a circle rearranged into a parallelogram by a method that has been known for hundreds of years. Figure 1(e) represents a different manipulation of area called *scaling* where, despite enlargement or shrinkage, shape and proportion are retained.

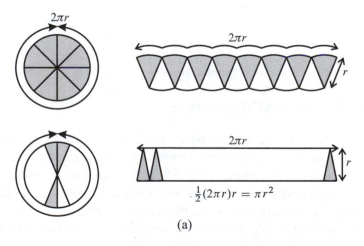

$$\tfrac{1}{2}(2\pi r)r = \pi r^2$$

(a)

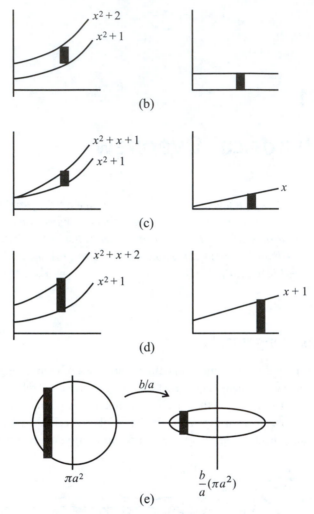

Figure 1. Examples of rearrangements (a–d) and scaling (e) of areas

1.2   The Lune of Hippocrates

Hippocrates (430 B.C.E.), a merchant of Athens, was one of the first to find the area of a plane figure (lune) bounded by curves (circular arcs). The crescent-shaped region whose area is to be determined is shown in Figure 2.

In the figure, ABC and AFC are circular arcs with centers E and D, respectively. Hippocrates showed that the area of the shaded region bounded by the circular arcs ABC and AFC is exactly the area of the shaded square

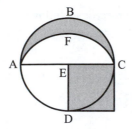

Figure 2. Hippocrates' lune

whose side is the radius of the circle. The argument depends on the following assumption, illustrated in Figure 3: *The areas of the two circles are to each other as the squares of the radii.*

Figure 3. Hippocrates' assumption

From this assumption we reach two conclusions:

1. The sectors of two circles with equal central angles are to each other as the squares of the radii (Figure 4).

Figure 4. Our first conclusion

2. The segments of two circles with equal central angles are to each other as the squares of the radii (Figure 5).

Figure 5. Our second conclusion

Hippocrates' argument proceeds as follows (refer to Figure 6). From our second conclusion, $A_1/A_4 = r^2/(\sqrt{2}r)^2 = \frac{1}{2}$. Hence $A_1 = \frac{1}{2}A_4$ and $A_2 = \frac{1}{2}A_4$ so $A_1 + A_2 = A_4$.

Area of the lune $= A_1 + A_2 + A_3 = A_4 + A_3 = $ Area of the triangle

$$= \frac{1}{2}\left(\sqrt{2}r\right)\left(\sqrt{2}r\right) = r^2 = \text{Area of the square}.$$

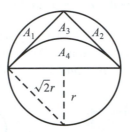

Figure 6. Area of the lune

Similar reasoning may be used for the three lunes in Figure 7.

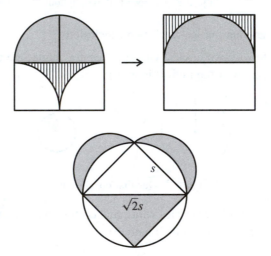

Figure 7. Lune exercises

1.3   Eudoxus and the Method of Exhaustion

Eudoxus (408–355 B.C.E.) was responsible for the notion of approximating curved regions with polygonal regions. In other words, "truth" for polygonal regions implies "truth" for curved regions. This notion will be used to show that the areas of circles are to each other as the squares of their diameters, an obvious result for regular polygons. "Truth" is to be based on Eudoxus' Axiom.

Axiom 1.3.1 (Axiom of Eudoxus). *Two unequal magnitudes being set out, if from the greater there be subtracted a magnitude greater than its half, and from that which is left a magnitude greater than its half, and if this process be repeated continuously, there will be left some magnitude that will be less than the lesser magnitude set out.*

In modern terminology, let M and $\epsilon > 0$ be given with $0 < \epsilon < M$. Then form M, $M - rM = (1-r)M$, $(1-r)M - r(1-r)M = (1-r)^2 M, \ldots$, where $\frac{1}{2} < r \le 1$. The axiom tells us that for n sufficiently large, say N, $(1-r)^N M < \epsilon$, a consequence of the set of natural numbers not being bounded above.

To get back to what we are trying to show, let c, C be circles with areas a, A and diameters d, D, respectively. We want to show that $a/A = d^2/D^2$, given that the result is true for polygons and given the Axiom of Eudoxus.

Assume that $a/A > d^2/D^2$. Then we have $a^* < a$, so that $0 < a - a^*$ and $a^*/A = d^2/D^2$. Let $\epsilon < a - a^*$. Inscribe regular polygons of areas p_n, P_n in circles c, C and consider the areas $a - p_n$, $A - P_n$. (See Figure 8.) Now, double the number of sides. What is the relationship between $a - p_n$ and $a - p_{2n}$? See Figure 9.

Certainly $a - p_{2n} < \frac{1}{2}(a - p_n)$. We subtract more than half each time we double the number of sides. From the Axiom of Eudoxus, we may determine N so that $0 < a - p_N < \epsilon < a - a^*$; that is, we have a regular inscribed polygon of N sides, where area $p_N > a^*$.

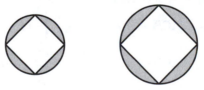

Figure 8. Inscribed polygons

$a - p_n$ $a - p_{2n}$

Figure 9. Segments

But $p_N/P_N = d^2/D^2$. Since $a^*/A = d^2/D^2$, we have $p_N/P_N = a^*/A$, so $P_N > A$. This cannot be: P_N is the area of a polygon inscribed in the circle C of area A.

A similar argument shows that a/A cannot be less than d^2/D^2: *double reductio ad absurdum*.

1.4 Archimedes' Method

The following masterpiece of mathematical reasoning is due to one of the greatest intellects of all time, Archimedes of Syracuse (287–212 B.C.E.). Archimedes shows that the area of the parabolic segment is $\frac{4}{3}$ that of the inscribed triangle ACB. See Figure 10.

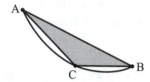

Figure 10. Archimedes' triangle

In our discussion, we will use the symbol $\triangle$ to denote "area of," with $x_C = (x_A + x_B)/2$. The argument proceeds as follows. In Figure 11, the combined area of triangles ADC and BEC is one-fourth the area of triangle ACB; that is,

$$\triangle ADC + \triangle BEC = \frac{1}{4}\triangle ACB.$$

Repeating the process, trying to "exhaust" the area between the parabolic curve and the inscribed triangles, we have

Area of the parabolic segment

$$= \triangle ACB + \frac{1}{4}(\triangle ACB) + \frac{1}{4}\left[\frac{1}{4}(\triangle ACB)\right] + \cdots$$

$$= \triangle ACB\left(1 + \frac{1}{4} + \frac{1}{4^2} + \cdots\right) = \frac{4}{3}\triangle ACB.$$

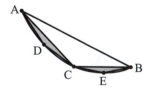

Figure 11. Inscribed triangles

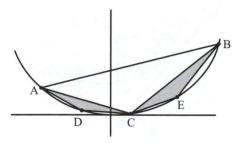

Figure 12.

We argue that $\triangle ADC + \triangle BEC = \frac{1}{4}\triangle ACB$ for the parabola $y = ax^2$, $a > 0$; see Figure 12.

Show that the tangent line at C is parallel to AB and that the vertical line through C bisects AB at P. It follows that $\triangle BEC = \frac{1}{4}\triangle BCP$ (complete the parallelogram; see Figure 13).

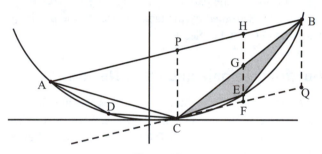

Figure 13.

We note that

1. $\triangle CEG = \triangle BEG$ (equal height and base),

2. $\triangle HGB = \frac{1}{4}\triangle BCP$.

Thus, we must show that

$$\triangle CEG + \triangle BEG = \triangle HGB \quad \text{or} \quad \triangle BEG = \frac{1}{2}\triangle HGB.$$

This will be accomplished by showing that $FE = \frac{1}{4}FH = \frac{1}{4}QB$. Since

$$FE = a\left(\frac{X_C + X_B}{2}\right)^2 - \left[aX_C^2 + 2aX_C \times \frac{1}{2}(X_B - X_C)\right]$$
$$= \frac{1}{4}a(X_B - X_C)^2,$$

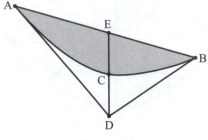

Figure 14.

and

$$QB = a X_B^2 - \left[a X_C^2 + 2 a X_C \left(X_B - X_C\right)\right] = a \left(X_B - X_C\right)^2,$$

we are done.

Show that the area of the parabolic segment is $\frac{2}{3}$ the area of the circumscribed triangle ADB formed by the tangent lines to the parabola at A and B with base AB (EC = CD). See Figure 14.

1.5   Gottfried Leibniz and Isaac Newton

During the seventeenth and eighteenth centuries the integral was thought of in a descriptive sense, as an *antiderivative*, because of the beautiful Fundamental Theorem of Calculus (FTC), as developed by Leibniz (1646–1716) and Newton (1642–1723).

A particular function f on $[a, b]$ was integrated by finding an antiderivative F so that $F' = f$, or by finding a power series expansion and using the FTC to integrate termwise. The Leibniz–Newton integral of f was $F(b) - F(a)$; that is,

$$\int_a^b f(x)\,dx = F(b) - F(a), \qquad \text{where } F' = f.$$

We will give an argument of Leibniz and a result of Newton to illustrate their geniuses.

1.5.1   Leibniz's Argument

Leibniz argued that

$$\frac{\pi}{4} = 1 - \frac{1}{3} + \frac{1}{5} - \frac{1}{7} + \cdots.$$

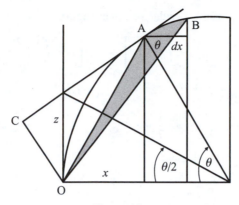

Figure 15.

Take the quarter circle $(x - 1)^2 + y^2 = 1$, $0 \le x \le 1$, whose area is $\pi/4$. (See Figure 15.) Leibniz determined the area of the circular sector in Figure 15 by dividing it into infinitesimal triangles OAB (where A and B are two close points on the circle) and summing.

So, how to estimate the area of $\triangle$OAB? Construct the tangent to the circle at A, with a perpendicular at C passing through the origin. Then $\triangle$OAB $\approx$ $\frac{1}{2}$AB $\times$ OC. By similar triangles, AB$/ dx = z/$OC, so $\triangle$OAB $= \frac{1}{2}z\, dx$. Observe that

$$x = 1 - \cos \theta = 2 \sin^2 \frac{\theta}{2} \qquad \text{and} \qquad z = \tan \frac{\theta}{2}.$$

That is, $x = 2x^2/(1 + z^2)$. Leibniz knew that

$$xz = \int z\, dx + \int x\, dz.$$

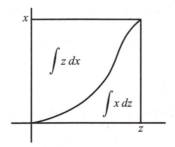

Figure 16.

Hence the area of the circular sector in Figure 15 is

$$\int_0^1 \frac{1}{2} z \, dx = \frac{1}{2} \left[xz \mid_0^1 - \int_0^1 x \, dz \right] = \frac{1}{2} \left[1 - \int_0^1 \frac{2z^2}{1 + z^2} \, dz \right]$$

$$= \frac{1}{2} - \int_0^1 z^2 (1 - z^2 + z^4 - \cdots) \, dz \qquad \text{(long division)}$$

$$= \frac{1}{2} - \frac{1}{3} + \frac{1}{5} - \frac{1}{7} + \cdots \qquad \text{(integrating termwise)}$$

and by adding $\frac{1}{2}$ to both sides,

$$\frac{\pi}{4} = 1 - \frac{1}{3} + \frac{1}{5} - \frac{1}{7} + \cdots .$$

By the way,

$$\frac{1}{8} \pi = \frac{1}{1 \times 3} + \frac{1}{5 \times 7} + \frac{1}{9 \times 11} + \cdots \quad \text{and}$$

$$\frac{1}{8} \ln 4 = \frac{1}{2 \times 4} + \frac{1}{6 \times 8} + \frac{1}{10 \times 12} + \cdots .$$

1.5.2 Newton's Result

Newton tells us that

$$\frac{\pi}{4\sqrt{2}} = 1 + \frac{1}{3} - \frac{1}{5} - \frac{1}{7} + \frac{1}{9} + \frac{1}{11} - \cdots .$$

Since Newton routinely integrated series termwise, and since

$$\frac{1 + x^2}{1 + x^4} = (1 + x^2)(1 - x^4 + x^8 - \cdots) = 1 + x^2 - x^4 - x^6 + \cdots ,$$

we have

$$\int_0^1 \frac{1 + x^2}{1 + x^4} \, dx = 1 + \frac{1}{3} - \frac{1}{5} - \frac{1}{7} + \frac{1}{9} + \frac{1}{11} - \cdots .$$

The argument may be completed by observing that

$$\frac{1 + x^2}{1 + x^4} = \frac{1}{2} \left(\frac{1}{1 - \sqrt{2}x + x^2} + \frac{1}{1 + \sqrt{2}x + x^2} \right)$$

and evaluating the appropriate integrals with the substitution

$$x + \frac{\sqrt{2}}{2} = \frac{1}{\sqrt{2}} \tan \theta.$$

1.6   Augustin-Louis Cauchy

Cauchy (1789–1857) is considered to be the founder of integration theory. In 1823 Cauchy formulated a constructive definition of an integral. Given a general function f on an interval $[a, b]$, in contrast to ax^2, $(1 + x^2)/(1 + x^4)$, and so on, partition the interval $[a, b]$ into subintervals $[x_{k-1}, x_k]$, with $a = x_0 < x_1 < \cdots < x_{n-1} < x_n = b$, and form the sum

$$f(x_0)(x_1 - x_0) + f(x_1)(x_2 - x_1) + \cdots + f(x_{n-1})(x_n - x_{n-1}).$$

See Figure 17.

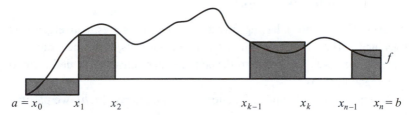

Figure 17. Cauchy's integral

The integral of Cauchy was to be the limit of such sums as the length of the largest subinterval, $\|\Delta x\|$, approaches zero:

$$C\int_a^b f(x)\,dx = \lim_{\|\Delta x\| \to 0} \sum_{k=1}^n f(x_{k-1})(x_k - x_{k-1}).$$

Cauchy argued that for continuous functions this limit always exists.

As for an evaluative procedure, recovering a function from its derivative (a fundamental result), we have

$$C\int_a^b F'(x)\,dx = F(b) - F(a)$$

for any function F with a continuous derivative. For example, let

$$F(x) = \begin{cases} x^3 \sin(\pi/x) & x \neq 0, \\ 0 & x = 0. \end{cases}$$

Then

$$F'(x) = \begin{cases} -\pi x \cos(\pi/x) + 3x^2 \sin(\pi/x) & x \neq 0, \\ 0 & x = 0 \end{cases}$$

is continuous on the interval $[0, 1]$. Consequently

$$\text{C} \int_0^1 F'(x)\, dx = F(1) - F(0) = 0.$$

Apparently a finite number of "jump" discontinuities would not cause difficulties. How about a countable number, or even a dense set, of jump discontinuities? Just how discontinuous can a function be and still have an integral?

1.7 Bernhard Riemann

Riemann (1826–1866), having investigated Fourier series, convergence issues, and Dirichlet-type functions (1 on the rationals, 0 on the irrationals, for example), was motivated to develop another constructive definition of an integral (1854).

Beginning with a bounded function f on the interval $[a, b]$, we partition (à la Cauchy) into subintervals $[x_{k-1}, x_k]$, where $a = x_0 < x_1 < \cdots < x_n = b$. Next we "tag" each subinterval with an arbitrary point c_k, where $x_{k-1} \leq c_k \leq x_k$, and we form the sum

$$f(c_1)(x_1 - x_0) + f(c_2)(x_2 - x_1) + \cdots + f(c_n)(x_n - x_{n-1}).$$

See Figure 18.

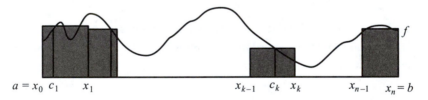

$a = x_0 \; c_1 \qquad x_1 \qquad\qquad\qquad\qquad x_{k-1} \; c_k \; x_k \qquad\qquad x_{n-1} \quad x_n = b$

Figure 18. Riemann's integral

Whereas the tag c_k was the left-hand endpoint in the Cauchy definition ($c_k = x_{k-1}$), in Riemann's definition we have more variability. Again, as the length of the largest subinterval approaches zero, the limit yields the Riemann integral:

$$\text{R} \int_a^b f(x)\, dx = \lim_{\|\Delta x\| \to 0} \sum_{k=1}^n f(c_k)(x_k - x_{k-1}).$$

In 1902 Lebesgue showed that for bounded functions, continuity is both necessary and sufficient for the existence of the Riemann integral (with the possible exception of a set of measure 0). For example, the function

$$f(x) = \begin{cases} 1/q & x = p/q \quad (p, q \text{ relatively prime natural numbers}), \\ 0 & \text{otherwise}, \end{cases}$$

is continuous on the irrationals, discontinuous on the rationals, and thus Riemann integrable. In fact, $R \int_0^1 f(x)\, dx = 0$. Riemann also constructed a function with a dense set of discontinuities that was Riemann integrable. (See Exercise 3.4.3.)

Every Cauchy integrable function is Riemann integrable and has the same value. We have a more general Fundamental Theorem of Calculus for recovering a function from its derivative.

Theorem 1.7.1 (General FTC for Riemann Integrability). *The integral*

$$R \int_a^b F'(x)\, dx = F(b) - F(a)$$

for any function with a derivative that is bounded and continuous almost everywhere.

For example,

$$F(x) = \begin{cases} x^2 \sin(\pi/x) & x \neq 0, \\ 0 & x = 0, \end{cases}$$

has the derivative

$$F'(x) = \begin{cases} -\pi \cos(\pi/x) + 2x \sin(\pi/x) & x \neq 0, \\ 0 & x = 0, \end{cases}$$

bounded and continuous except at $x = 0$. Thus, the Riemann integral of F' exists and

$$R \int_0^1 F'(x)\, dx = F(1) - F(0) = 0.$$

In 1881 Vito Volterra gave an example of a differentiable function with a bounded derivative that was discontinuous on a set of positive measure and thus not Riemann integrable. (See Section 3.12.) Again, the function $x^2 \sin(\pi/x)$, modified on a Cantor set of positive measure, sufficed.

This example prompted Lebesgue to develop an integral to remedy this defect. It turns out that

$$L \int_a^b F'(x)\, dx = F(b) - F(a)$$

for a differentiable function F with a bounded derivative. More about this later.

1.8 Thomas Stieltjes

Stieltjes (1856–1894) was interested in mathematically modelling mass distributions on the real line. Suppose we have point masses distributed as indicated in Figure 19. If $\phi(x)$ denotes the total mass less than or equal to x, then the graph of ϕ appears as shown in Figure 20. In general, ϕ is a nondecreasing function.

Figure 19. Masses on the real line

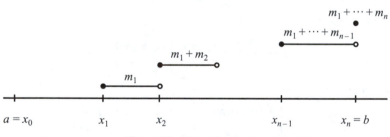

Figure 20. Mass distribution

Now consider the moment of such a mass distribution. The "mass" of $[x_{k-1}, x_k]$ is $\phi(x_k) - \phi(x_{k-1})$ with "arm" c_k, for $x_{k-1} \leq c_k \leq x_k$. This leads to sums of the form

$$c_1 m_1 + c_2 m_2 + \cdots + c_n m_n \qquad \text{or}$$
$$c_1[\phi(x_1) - \phi(x_0)] + \cdots + c_n[\phi(x_n) - \phi(x_{n-1})].$$

More generally, Stieltjes was led to consider sums — *Riemann–Stieltjes sums* — like $f(c_1)m_1 + \cdots + f(c_n)m_n$. We have a "weighted" sum. The value of the function at c_1, $f(c_1)$ is weighted by $m_1; \ldots;$ the value of the function at c_n, $f(c_n)$, is weighted by m_n. The average would be

$$\frac{\sum f(c_k) m_k}{\sum m_k} = \frac{\sum f(c_k)[\phi(x_k) - \phi(x_{k-1})]}{[\phi(b) - \phi(a)]}$$

$$\approx \frac{\sum f(c_k)\phi'(\xi_k)(x_k - x_{k-1})}{\int_a^b \phi'(x)\,dx}$$

$$\approx \frac{\int_a^b f(x)\phi'(x)\,dx}{\int_a^b \phi'(x)\,dx}$$

for "nice" f and ϕ.

What conditions may we impose on f and ϕ to make these suggestive manipulations legitimate? Heuristics has suggested

$$\text{R-S} \int_a^b f(x)\,d\phi(x) = \text{R} \int_a^b f(x)\phi'(x)\,dx.$$

The Riemann integral makes sense for functions that are bounded and continuous almost everywhere. So, f continuous and ϕ' Riemann integrable should work — and it does: see Theorem 4.3.1.

Do we have anything new here? Formally,

$$\text{R-S} \int_a^b f(x)\,d\phi(x) \equiv \lim_{\|\Delta x\| \to 0} \sum_1^n f(c_k)[\phi(x_k) - \phi(x_{k-1})].$$

We can show this limit makes sense for f continuous and ϕ monotone (Theorem 4.4.1). Of course, functions of bounded variation are differences of two monotone functions, so it is true for ϕ a function of bounded variation.

Another question: Does the series $\sum \cos(\sqrt{n})/n$ converge? This is an amusing application of the Euler Summation Formula (Section 4.5). The Riemann–Stieltjes integral is very convenient for step functions.

By the way, evaluate $\text{R} \int_{k-1}^k [x - (k - 1) - \frac{1}{2}] f'(x)\,dx$ by parts (for $2 \leq k \leq n$), sum the results, and note that $x - [x] - \frac{1}{2}$ differs from $x - (k - 1) - \frac{1}{2}$ at a finite number of points. Try $f(x) = \ln x$ for a "Stirling" result.

1.9   Henri Lebesgue

Where to begin? The Lebesgue integral has affected many areas of mathematics during the past century. Let's begin as Henri Lebesgue (1875–1941) did, with Volterra's example of a function with a bounded derivative that was not Riemann integrable (see Section 3.12), for it was this example that prompted Lebesgue to develop an integral (1902) that would recover any function from its bounded derivative. That is, $L \int_0^1 F'(x)\,dx = F(b) - F(a)$ should be true whenever the derivative F' is bounded.

Lebesgue's integral construction was fundamentally different from his predecessors. His simple, but brilliant, idea was to partition the *range* of the function rather than its domain.

Assume $\alpha < f < \beta$ on the interval $[a, b]$. In place of $a = x_0 < x_1 < \cdots < x_n = b$, we have $\alpha = y_0 < y_1 < \cdots < y_n = \beta$. The sets

$$f^{-1}([y_{k-1}, y_k)) = \{x \in [a, b] \mid y_{k-1} \le f(x) < y_k\}$$

are disjoint with union $[a, b]$. Disregarding the empty sets (relabelling if necessary), pick a tag (point c_k) in each nonempty set, and form the sum (motivated by areas of rectangles as the height times the length of the base) as follows (see Figure 21):

$$f(c_1)\cdot\{\text{length of } f^{-1}([y_0, y_1))\} + \cdots + f(c_n)\cdot\{\text{length of } f^{-1}([y_{n-1}, y_n))\}.$$

We then have

$$\sum y_{k-1} \cdot \{\text{length of } f^{-1}([y_{k-1}, y_k))\}$$
$$\le \sum f(c_k) \cdot \{\text{length of } f^{-1}([y_{k-1}, y_k))\}$$
$$\le \sum y_k \cdot \{\text{length of } f^{-1}([y_{k-1}, y_k))\}$$

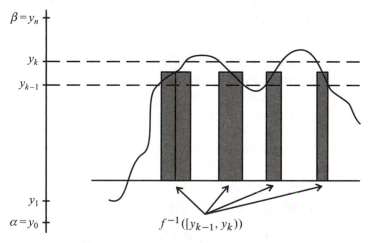

Figure 21. Lebesgue's integral construction

and

$$\sum (y_k - y_{k-1}) \cdot \{\text{length of } f^{-1}([y_{k-1}, y_k))\} \leq \|\Delta y\|(b - a).$$

What do we mean by the "length" of $f^{-1}([y_{k-1}, y_k))$? Partitioning the range forces us to assign a length, or measure, to possibly unusual sets.

For example, suppose we are dealing with the Dirichlet function on the interval $[0, 1]$ (assign a functional value of 1 whenever x is irrational and a value of -1 whenever x is rational). A partitioning of the range, say $-1\frac{1}{2}, -\frac{1}{2}, \frac{1}{2}, 1\frac{1}{2}$, would compel us to assign a length to the sets of rational, $f^{-1}([-1\frac{1}{2}, -\frac{1}{2}))$, and irrational $f^{-1}([\frac{1}{2}, 1\frac{1}{2}))$, numbers in the interval $[0, 1]$. Both sets, being subsets of $[0, 1]$, should have length less than or equal to 1. Since their union is the interval $[0, 1]$, the sum of their lengths should be 1.

Let's see, we could enumerate the rationals, $r_1, r_2, \ldots, r_n, \ldots$, and cover each rational with an interval $(r_n - \epsilon/2^n, r_n + \epsilon/2^n)$, which covers all the rationals with an open set of length less than ϵ. Fine; the rationals will have length 0, the irrationals will have length 1. It happens that this Dirichlet function is Lebesgue integrable (not Riemann integrable) and

$$L \int_0^1 f(x) \, dx = \left[1 \cdot (\text{length of irrationals in } [0, 1]) \right]$$
$$+ \left[-1 \cdot (\text{length of rationals in } [0, 1]) \right]$$
$$= 1.$$

What should the "length" of the numbers in $[0, 1]$ without a 5 in their decimal expansion be? How about the Cantor set? Removing intervals would suggest that the Cantor set, even though uncountable, has measure 0.

A "measure" theory must be developed that is logical and consistent. This integral of Lebesgue, if it is to have any power, suggests that we should be able to measure most sets of real numbers. Through the efforts of Jordan, Borel, Lebesgue, and Carathéodory, to name a few, a nonnegative countably additive measure was developed — the *Lebesgue measure* — that measures, in particular, all Borel sets of real numbers (countable unions of countable intersections of ... open sets).

In short, nonmeasurable sets are difficult to construct, so sets without "length" will seldom occur.

So, we want $f^{-1}([y_{k-1}, y_k))$ to be a measurable set. We want inverse images of intervals to be measurable. We therefore define a *measurable function* to be a function for which inverse images of intervals are measurable. Again, "most" functions are measurable, and because the limit of a sequence of measurable functions is measurable, we have some beautiful convergence theorems.

For example, enumerate the rationals in the interval $[0, 1]$, and define a sequence of measurable and Riemann integrable functions $\{f_k\}$ by $f_k(x) = 1$, $x = r_1, r_2, \ldots, r_k$, and $f_k(x) = 0$ otherwise. This sequence is nonnegative, monotone increasing, uniformly bounded by 1; $\lim f_k$ is a Dirichlet function, and thus it is not Riemann integrable: clearly $\lim R \int_a^b f_k(x)\, dx = 0$, and $R \int_a^b \lim f_k(x)\, dx$ is not defined, but

$$\mathrm{L}\int_a^b f_k(x)\, dx = 0 = \mathrm{L}\int_a^b \lim f_k(x)\, dx.$$

Finally, in answer to Volterra, all functions with a bounded derivative are Lebesgue integrable and $\mathrm{L}\int_0^1 F'(x)\, dx = F(b) - F(a)$. (See Theorem 6.4.2.)

1.10   The Lebesgue–Stieltjes Integral

The construction of Lebesgue measure begins with the assignment of a measure to an interval, namely its length: the measure of $(a, b]$ is $b - a$. Just as in the Riemann–Stieltjes integral, where we weighted the interval $(a, b]$ by $\phi(b) - \phi(a)$, a particularly fruitful approach to the construction of Lebesgue–Stieltjes measure is to assign a measure of $\phi(b) - \phi(a)$ to the interval $(a, b]$ where ϕ is a nonnegative, monotone increasing, right-continuous function on the reals, with

$$\lim_{x \to -\infty} \phi(x) = 0, \qquad \lim_{x \to +\infty} \phi(x) = 1.$$

It turns out, just as with ordinary Lebesgue measure, that a nonnegative, countably additive measure, μ_ϕ, is generated on the Borel sets of real numbers. From this so-called *Lebesgue–Stieltjes measure*, we proceed to measurable functions and Lebesgue–Stieltjes integrals.

For example, if

$$\phi(x) = \begin{cases} 0 & x < 0, \\ x^2 & 0 \le x \le 1, \\ 1 & 1 < x, \end{cases}$$

and if f is the Dirichlet function, what is L–S $\int_R f(x)d\mu_\phi$? It would be helpful if

$$\text{L–S} \int_R f(x)d\mu_\phi = \text{L} \int_0^1 f(x)(x^2)'d\mu$$

$$= \text{L} \int_0^1 f(x)2x dx$$

$$= \text{L} \int_{\text{irrationals}} 1 \cdot 2x dx + \text{L} \int_{\text{rationals}} 0 \cdot 2x dx$$

$$= \text{L} \int_0^1 1 \cdot 2x dx = 1.$$

(See Theorem 7.7.1.)

As it happens, the Lebesgue–Stieltjes integral is crucial in probability.

1.11 Ralph Henstock and Jaroslav Kurzweil

Working independently, Henstock (1923–) and Kurzweil (1926–) discovered the generalized Riemann integral in 1961 and 1957, respectively. Their discovery, which is referred to as the H-K integral, is an extension of the Lebesgue integral. All Lebesgue integrable functions are H-K integrable functions, to the same value. What's more, there are H-K integrable functions that are not Lebesgue integrable.

If a function is Lebesgue integrable then its absolute value must be Lebesgue integrable.

Consider the function

$$F(x) = \begin{cases} x^2 \sin(\pi/x^2) & x \neq 0, \\ 0 & x = 0, \end{cases}$$

and its derivative

$$F'(x) = \begin{cases} -2\pi/x \cos(\pi/x^2) + 2x \sin(\pi/x^2) & x \neq 0, \\ 0 & x = 0. \end{cases}$$

The Lebesgue integral of $|F'|$ does not exist.

Consider the intervals $\left[\sqrt{2/(4k+3)}, \sqrt{2/(4k+1)} \right]$:

$$L \int_0^1 |F'(x)| \, dx$$

$$\geq \sum_1^n \int_{a_k}^{b_k} |F'(x)| \, dx \geq \sum_1^n \left| \int_{a_k}^{b_k} F'(x) \, dx \right|$$

$$= \sum_1^n |F(b_k) - F(a_k)|$$

$$= \sum_1^n \left[\frac{2}{4k+1} \sin\left(\frac{4k+1}{2} \right) \pi - \frac{2}{4k+3} \sin\left(\frac{4k+3}{2} \right) \pi \right]$$

$$\geq \sum_1^n \left[\left(\frac{2}{4k+1} \right) (1) - \left(\frac{2}{4k+3} \right) (-1) \right]$$

$$\geq \sum_1^n \frac{1}{k+1}.$$

It turns out that every derivative is H-K integrable.

Thus in this example H-K $\int_0^1 F'(x) \, dx = F(1) - F(0) = 0$. (Think about the graph of F'.)

This very powerful integral results from an apparently simple modification of the Riemann integral construction. Rather than partitioning the interval $[a, b]$ into a collection of subintervals of fairly uniform length, and then selecting a tag (point) c_k from each subinterval at which to evaluate the function, we will be guided by the behavior of the function in the assignment of a subinterval. If the function oscillates, or behaves unpleasantly about a point c, we associate a small subinterval with c. If the function is better behaved, we associate a larger subinterval.

With the Riemann integral, to obtain accurate approximations by sums of the form $f(c_1)(x_1 - x_0) + \cdots + f(c_n)(x_n - x_{n-1})$, we required the maximum lengths of the subintervals, $\|\Delta x\|$, to be less than some constant δ. With the H-K integral, however, the δ that regulates lengths of subintervals will be a function. A subinterval $[u, v]$ with a tag c must satisfy $c - \delta(c) < u \leq c \leq v < c + \delta(c)$. A partition of $[a, b]$ will be determined by a positive function $\delta(\cdot)$ so that $a = x_0 < x_1 < \cdots < x_n = b$, with the requirement that $c_k - \delta(c_k) < x_{k-1} \leq c_k \leq x_k < c_k + \delta(c_k)$.

The H-K sums exhibit the same appearance as the ordinary Riemann sums $f(c_1)(x_1 - x_0) + \cdots + f(c_n)(x_n - x_{n-1})$, but with the H-K integral δ is a positive function on $[a, b]$, where $x_k - x_{k-1} < 2\delta(c_k)$, $x_{k-1} \leq c_k \leq x_k$.

Example 1.11.1. For an example, let's begin with the Lebesgue integrable Dirichlet function on the interval $[0, 1]$ that is 1 on the rationals and 0 on the irrationals.

Consider any Riemann sum $\sum f(c_k)(x_k - x_{k-1})$. There will be no contribution to this sum unless the tag c_k is a rational number. We want any interval associated with such a tag to be "small." Enumerate the rationals in $[0, 1]$: $r_1, r_2, \ldots, r_n, \ldots$. Define a positive function $\delta(\cdot)$ on $[0, 1]$ by

$$\delta(c) = \begin{cases} \epsilon & c = r_1, r_2, \ldots, r_n, \ldots, \\ 1 & \text{otherwise.} \end{cases}$$

Then any Riemann sum is nonnegative and $\sum f(c_k)(x_k - x_{k-1}) \leq \sum_1^\infty 1 \cdot 2\epsilon$.

We want convergence here. Redefine

$$\delta(c) = \begin{cases} \epsilon/2^{n+1} & \epsilon = r_n, \\ 1 & \text{otherwise.} \end{cases}$$

Then we may conclude

$$\sum f(c_k)(x_k - x_{k-1}) \leq \sum_1^\infty 1 \cdot \frac{2\epsilon}{2^{n+1}} = \epsilon.$$

We have glossed over two difficulties. First, just because we have a positive function $\delta(\cdot)$ on the interval $[a, b]$, how do we know there is a partition of $[a, b]$ so that $c_k - \delta(c_k) < x_{k-1} \leq c_k \leq x_k < c_k + \delta(c_k)$? (This was settled by Cousin in 1885.)

Second, to use this integral effectively, we need to be able to construct suitable positive functions $\delta(\cdot)$ for a particular function f, from the vague idea that erratic behavior of the function at a point generally requires small subintervals about that point.

But in the end we are rewarded handsomely: we have better "Fundamental Theorems," better convergence theorems, and so on, than we have with the Lebesgue integral. Whereas the Lebesgue integral was the integral of the twentieth century, the H-K integral may lead to new developments in the twenty-first century. In fact, P. Muldowney (1987) treats two of the integrals we discuss later — the Wiener integral and the Feynman integral — as special cases of the H-K integral over function spaces.

On the other hand, the Lebesgue integral is particularly suited for L^p spaces. Recall L. Carleson's result that the Fourier series of an L^2 function converges almost everywhere.

1.12   Norbert Wiener

The *Wiener integral*, developed by Norbert Wiener (1894–1964) in the
1920s, was a spectacular advance in the theory of integration. Wiener constructed a measure on a function space — in fact, the Banach space of
continuous functions — on the interval $[0, 1]$ beginning at the origin, with
$\|x\| = \sup_{0 \le t \le 1} |x(t)|$. See Figure 22.

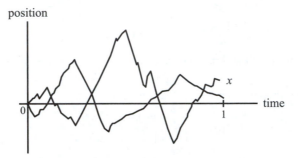

Figure 22. Wiener's continuous functions

This measure arose when Wiener was trying to understand Brownian
motion (e.g., that of pollen grains in suspension, moving erratically). Think
of a collection of particles at position $x(0) = 0$ at time 0, moving to position
$x(1)$ at time 1 (see Figure 23).

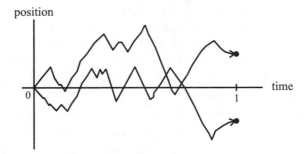

Figure 23. Brownian motion

Now suppose $0 < t_1 \le 1$ and $a_1 < x(t_1) \le b_1$. We may think of the set
of continuous functions (particles) on $[0, 1]$ that pass through the "window"
$(a_1, b_1]$ at time t_1, a *quasi-interval* in Wiener's terms (see Figure 24).

We want to measure the fraction of the particles that begin at position 0
at time 0 and pass through the window $(a_1, b_1]$ at time t_1. From physical

position x

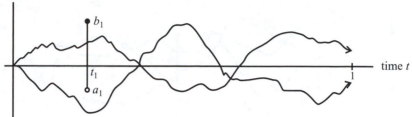

Figure 24. Quasi-interval

considerations and genius, Wiener assigned a measure, w:

$$w\Big(\{x(\cdot) \in C_0 \mid a_1 < x(t_1) \le b_1, 0 < t_1 \le 1\}\Big)$$

$$= L \int_{a_1}^{b_1} (2\pi t_1)^{-1/2} e^{-\xi_1^2/2t_1} d\xi_1.$$

We have, for example,

$$\phi = \big(\{x(\cdot) \in C_0 \mid 0 < x(t_1) \le 0, 0 < t_1 \le 1\}\big) \quad \text{and}$$

$$w(\phi) = L \int_0^0 (2\pi t_1)^{-1/2} e^{-\xi_1^2/2t_1} d\xi_1 = 0.$$

Certainly all the particles will pass through the large window $(-\infty, \infty]$ at t_1. So,

$$w(C_0) = w\Big(\{x(\cdot) \in C_0 \mid -\infty < x(t_1) \le \infty, \ 0 < t_1 \le 1\}\Big)$$

$$= L \int_{-\infty}^{\infty} (2\pi t_1)^{-1/2} e^{-\xi_1^2/2t_1} d\xi_1 = 1.$$

Now, suppose we have two windows, $(a_1, b_1]$ at t_1 and $(a_2, b_2]$ at t_2, where $0 < t_1 < t_2 \le 1$. See Figure 25.

We have

$$\{x(\cdot) \in C_0 \mid a_1 < x(t_1) \le b_1, a_2 < x(t_2) \le b_2, 0 < t_1 < t_2 \le 1\}.$$

Wiener assigned a measure of

$$L \int_{a_2}^{b_2} d\xi_2 \int_{a_1}^{b_1} d\xi_1 (2\pi t_1)^{-\frac{1}{2}} e^{-\xi_1^2/2t_1} [2\pi(t_2 - t_1)]^{-\frac{1}{2}} e^{-(\xi_2 - \xi_1)^2/2(t_2 - t_1)}.$$

If the window at time t_1 is large, $(-\infty, \infty]$, no real restriction is imposed on the number of particles, and the measure of

$$\{x(\cdot) \in C_0 \mid -\infty < x(t_1) \le \infty, a_2 < x(t_2) \le b_2\}$$

position x

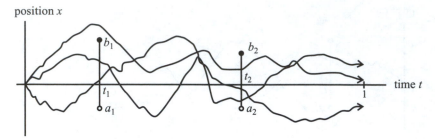

Figure 25.

should be the same as the measure of $\{x(\cdot) \in C_0 \mid a_2 < x(t_2) \leq b_2\}$. Show that

$$
\text{L} \int_{a_2}^{b_2} d\xi_2 \int_{-\infty}^{\infty} d\xi_1 (2\pi t_1)^{-1/2}
$$

$$
\cdot e^{-\xi_1^2/2(t_2-t_1)} [2\pi(t_2 - t_1)]^{-1/2} e^{-(\xi_2-\xi_1)^2/2(t_2-t_2)}
$$

$$
= \text{L} \int_{a_2}^{b_2} d\xi_2 (2\pi t_2)^{-1/2} e^{-\xi_2^2/2t_2}.
$$

Similarly if $(a_2, b_2] = (-\infty, \infty]$. Also,

$$
w\big(\{x(\cdot) \in C_0 \mid -\infty < x(t_1) \leq \infty, -\infty < x(t_2) \leq \infty, 0 < t_1 < t_2 \leq 1\}\big)
$$

$$
= 1.
$$

Finally, let $K(x, t) = (2\pi t)^{-1/2} e^{-x^2/2t}$, with $0 < t_1 < t_2 < \cdots < t_n \leq 1$. See Figure 26. Consider

$$
\{x(\cdot) \in C_0 \mid a_k < x(t_k) \leq b_k, 1 \leq k \leq n, 0 < t_1 < t_2 < \cdots < t_n \leq 1\}.
$$

position x

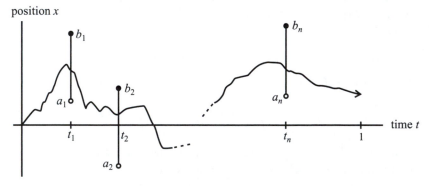

Figure 26. The n-dimensional quasi-interval

This quasi-interval will be assigned a measure,

$$w\big(\{x(\cdot) \in C_0 | a_k < x(t_k) \le b_k, \, 1 \le k \le n, \, 0 < t_1 < \cdots < t_2 \le 1\}\big)$$

$$= L \int_{a_n}^{b_n} d\,\xi_n \cdots \int_{a_1}^{b_1} d\,\xi_1 K(\xi_1, t_1) K(\xi_2 - \xi_1, t_2 - t_1)$$

$$\cdots K(\xi_n - \xi_{n-1}, t_n - t_{n-1}).$$

Wiener was able to extend this measure w on the quasi-intervals (finite number of restrictions) to a measure μ_w on the sigma-algebra generated by the quasi-intervals. For example, if $S = \{x(\cdot) \in C_0 \mid -1 \le x(t) \le 1, 0 \le t \le 1\}$, let

$$S_1 = \{x(\cdot) \in C_0 \mid -2 < x(1) \le 1\},$$
$$S_2 = \big\{x(\cdot) \in C_0 \mid -\tfrac{3}{2} < x\left(\tfrac{1}{2}\right), x(1) \le 1\big\},$$
$$\vdots$$
$$S_n = \left\{x(\cdot) \in C_0 \mid -1 - \frac{1}{n} < x\left(\frac{k}{2^{n-1}}\right) \le 1, \, k = 1, 2, \ldots, 2^{n-1}\right\}.$$

Then $S_1 \supset S_2 \supset \cdots \supset S_n \supset \cdots$ and $S = \cap S_n$. The set S is thus measurable as a countable intersection of quasi-intervals.

From the measure μ_w we have measurable functionals and, finally, Wiener integrals — "path" integrals. For example, suppose we have the functional $F[x(\cdot)] = x(t_0)$; that is, to each element $x(\cdot)$ of the function space C_0 we assign its value at $t = t_0, x(t_0)$. What should $\int_{C_0} F[x(\cdot)] d\mu_w$ be?

Formally,

$$\int_{C_0} x(t_0) d\mu_w = \int_{-\infty}^{\infty} \xi (2\pi t_0)^{-1/2} e^{-\xi^2/2t_0} d\xi = 0.$$

The expected value of its position at any t, for $0 < t \le 1$, should be 0.

Suppose $F[x(\cdot)] = x^2(t_0)$ for $0 < t_0 \le 1$. Formally,

$$\int_{C_0} x^2(t_0) d\mu_w = \int_{-\infty}^{\infty} \xi^2 (2\pi t_0)^{-1/2} e^{-\xi^2/2t_0} \, d\xi = t_0.$$

The variance is t for $0 < t \le 1$. It would be convenient if $F[x(\cdot)] = L \int_0^1 x^2(\tau) d\tau$, and

$$\int_{C_0} F[x(\cdot)] d\mu_w = \int_{C_0} \left(L \int_0^1 x^2(\tau) \, d\tau\right) d\mu_w$$

$$= \int_0^1 \left(\int_{C_0} x^2(\tau) \, d\mu_w\right) d\tau = L \int_0^1 \tau \, d\tau = \frac{1}{2}.$$

Such is the case.

1.13   Richard Feynman

Consider a quantum mechanical system Ψ that satisfies Schrödinger's equation,

$$\frac{\partial \Psi}{\partial t} = \frac{i\hbar}{2m}\frac{\partial^2 \Psi}{\partial t^2} - \frac{i}{\hbar}V, \qquad \text{for } -\infty < x < \infty, \ t > 0$$

with $\psi(0, x) = f(x)$, $-\infty < x < \infty$, and $\int_{-\infty}^{\infty} |\Psi|^2\, dx = 1$ for all $t \geq 0$. To explain the evolution in time of Ψ, Feynman (1918–1988) developed an integral interpretation of Ψ as a limit of Riemann-type sums (1948).

It was understood that $|\Psi|^2$ is a probability density function, and Feynman concluded that the total probability amplitude Ψ is the sum, over all continuous paths from position x_0 at time 0 to position x at time t, of the individual probability amplitudes. That is,

$$\Psi = \sum_{\substack{\text{all connecting} \\ \text{continuous paths}}} K e^{i/\hbar A[x(\cdot)]},$$

with $A[x(\cdot)] = \int_0^t \left[\frac{1}{2}\dot{m}x^2(\tau) - V\big(x(\tau)\big)\right] d\tau$ and K a normalizing constant. His idea was to approximate this expression with Riemann sums and take the limit. Thus,

$$\sum_{\substack{\text{all connecting} \\ \text{continuous paths}}} K e^{i/\hbar A[x]}$$

$$\approx \lim_{n\to\infty} K \int_R dx_{n-1} \cdots \int_R dx_0$$

$$\exp\left\{ \frac{\xi}{2\hbar(t/n)} \sum_1^n (x_k - x_{k-1})^2 - \frac{i}{\hbar}\left(\frac{t}{n}\right)\sum_1^n V(x_{k-1}) \right\} f(x_0),$$

where

$$K = \left(\frac{m}{2\pi i\hbar(t/n)}\right)^{n/2} \qquad \text{and} \qquad x_n = x.$$

We have an integral over a function space, $C_0[0, 1]$, the same function space as the Wiener integral. The integrand is different, however. The exponential term has modulus 1 and becomes highly oscillatory as $n \to \infty$. Furthermore, for the normalizing constant K, $|K| \to \infty$ as $n \to \infty$.

Interpretation of this limit, and an explanation of the convergence issues involved, has occupied physicists and mathematicians for over fifty years. In the chapter on the Feynman integral, we will see a brilliant explanation due to Edward Nelson, discovered in the 1960s.

1.14   References

1. Carleson, Lennart. On convergence and growth of partial sums of Fourier series. *Acta Mathematica* 116 (1966) 135–157.

2. Muldowney, Patrick. *A General Theory of Integration in Function Spaces.* Essex: Longman Scientific and Technical, and New York: Wiley, 1987.

3. Nelson, Edward. Feynman integrals and the Schrödinger equation. *Journal of Mathematical Physics* 5(1964) 332–43.

CHAPTER 2

The Cauchy Integral

The sole aim of science is the honor of the human mind, and from this point of view a question about numbers is as important as a question about the system of the world. — C. G. J. Jacobi

2.1 Exploring Integration

Augustin-Louis Cauchy (1789–1857) was the founder of integration theory. Before Cauchy, the emphasis was on calculating integrals of specific functions. For example, in our calculus courses we use the formulae

$$1 + 2 + \cdots + n = \frac{n(n+1)}{2} \qquad \text{and}$$

$$1^2 + 2^2 + \cdots + n^2 = \frac{n(n+1)(2n+1)}{6}$$

to show, using approximation by interior and exterior rectangles, that the areas of the regions between the x-axis and the curves $y = x$ and $y = x^2$ for $0 \le x \le b$ are given by $\int_0^b x \, dx = b^2/2$ and $\int_0^b x^2 \, dx = b^3/3$ — results obtained much earlier by Archimedes (287–212 B.C.E).

Such beautiful results begged for extension, most accomplished by sheer ingenuity. Here are some examples, formulae owing to Fermat, Wallis, Stirling, and Stieltjes.

2.1.1 Fermat's Formula

From Pierre de Fermat (1601–1665) we have

$$\int_0^b x^q dx = \frac{b^{q+1}}{q+1}, \qquad \text{for } q \text{ a positive rational number, } b > 0.$$

This formula may be justified as follows.

Divide $[0, b]$ into an infinite sequence of subintervals of varying widths with endpoints br^n, $0 < r < 1$, $n = 0, 1, \ldots$, and erect a rectangle of height $(br^n)^q$ over the subinterval $[br^{n+1}, br^n]$. Sum the areas of the "exterior" rectangles, and show that this sum is $b^{q+1}(1-r)/(1-r^{q+1})$. Evaluate the limit of this sum as r approaches 1.

Exercise 2.1.1. Show

$$\int_0^\pi \sin(x)\,dx = 2,$$

using the familiar limit, $\lim_{\theta \to 0}(\sin\theta)/\theta = 1$, and Lagrange's identity,

$$\sin\left(\frac{\pi}{n}\right) + \sin\left(\frac{2\pi}{n}\right) + \cdots + \sin\left(\frac{n\pi}{n}\right)$$

$$= \frac{\cos(\pi/2n) - \cos([(2n+1)\pi]/2n)}{2\sin(\pi/2n)}.$$

2.1.2 Wallis's Formula

John Wallis (1616–1703) gave us this formula:

$$\frac{\pi}{2} = \frac{2}{1} \cdot \frac{2}{3} \cdot \frac{4}{3} \cdot \frac{4}{5} \cdot \ldots \cdot \frac{2n}{2n-1} \cdot \frac{2n}{2n+1} \cdot \ldots .$$

Using integration by parts, show that

$$\int_0^\pi \sin^{n+2}(x)\,dx = \frac{n+1}{n+2} \int_0^\pi \sin^n(x)\,dx.$$

Next, since $\int_0^\pi \sin^{2n+2}(x)\,dx \le \int_0^\pi \sin^{2n+1}(x)\,dx \le \int_0^\pi \sin^{2n}(x)\,dx$, we conclude

$$\frac{2n+1}{2n+2} \le \frac{\int_0^\pi \sin^{2n+1}(x)\,dx}{\int_0^\pi \sin^{2n}(x)\,dx} \le 1.$$

Show

$$\int_0^\pi \sin^{2n}(x)\,dx = \pi \left(\frac{1}{2}\right)\left(\frac{3}{4}\right)\cdots\left(\frac{2n-1}{2n}\right) \qquad \text{and}$$

$$\int_0^\pi \sin^{2n+1}(x)\,dx = 2 \left(\frac{2}{3}\right)\left(\frac{4}{5}\right)\cdots\left(\frac{2n}{2n+1}\right).$$

Substituting terms, we have

$$\frac{\pi}{2} \cdot \frac{\int_0^\pi \sin^{2n+1}(x)\,dx}{\int_0^\pi \sin^{2n}(x)\,dx} = \frac{2}{1} \cdot \frac{2}{3} \cdot \frac{4}{3} \cdot \frac{4}{5} \cdot \ldots \cdot \frac{2n}{2n-1} \cdot \frac{2n}{2n+1}.$$

Taking the limit as $n \to \infty$, the result follows.

It is interesting to note that

$$
\frac{\pi^2}{8} = \int_0^{\pi/2} x \, dx = \int_0^{\pi/2} \left[\sin^{-1}\left(\sin(x) \right) \right] dx
$$

$$
= \int_0^{\pi/2} \left[\sin(x) + \left(\frac{1}{2} \right) \frac{\sin^3(x)}{3} + \left(\frac{1}{2} \right)\left(\frac{3}{4} \right) \frac{\sin^5(x)}{5} + \cdots \right]
$$

$$
= 1 + \frac{1}{3^2} + \frac{1}{5^2} + \cdots .
$$

2.1.3 Stirling's Formula

James Stirling (1692–1770) provides the formula

$$
\frac{n!}{\sqrt{2\pi n}(n/e)^n} \to 1 \qquad \text{as } n \to \infty.
$$

Let $f(x) = \ln x$, where $1 \le x \le n$. Using the trapezoidal rule, show that

$$
\int_1^n \ln x \, dx - \frac{1}{2}\{(\ln 1 + \ln 2) + (\ln 2 + \ln 3) + \cdots + [\ln(n-1) + \ln n]\}
$$

$$
= \text{"error terms."}
$$

Because $\int_1^n \ln x \, dx = n \ln n - n + 1$, we have $\ln(n!) = (n + \frac{1}{2}) \ln n - n + E_n$, where E_n has a limit as $n \to \infty$. Conclude that

$$
s_n = \frac{n!}{\sqrt{n}(n/e)^n} \to C, \qquad \text{as } n \to \infty, \, C > 0.
$$

Then by Wallis's formula, we have

$$
\frac{s_n^2}{s_{2n}} = \sqrt{2}\, \frac{2^{2n}(n!)^2}{\sqrt{n}(2n)!} \to \sqrt{2\pi}.
$$

Thus $n! / \left[\sqrt{2\pi n}(n/e)^n \right] \to 1$ as $n \to \infty$.

2.1.4 Stieltjes' Formula

From Thomas Stieltjes (1856–1894) we have that

$$
\int_0^\infty e^{-x^2} \, dx = \frac{\sqrt{\pi}}{2}.
$$

Using integration by parts, show that

$$\int_0^\infty x^{n+2} e^{-x^2} \, dx = \frac{n+1}{2} \int_0^\infty x^n e^{-x^2} \, dx.$$

Thus,

$$\int_0^\infty x^{2n} e^{-x^2} \, dx = \frac{1 \cdot 3 \cdot \cdots \cdot (2n-1)}{2^n} \int_0^\infty e^{-x^2} \, dx \qquad \text{and}$$

$$\int_0^\infty x^{2n+1} e^{-x^2} \, dx = \frac{1 \cdot 2 \cdot \cdots \cdot n}{2}.$$

Given $I_k = \int_0^\infty x^k e^{-x^2} \, dx$, we have

$$\alpha^2 I_{k-1} + 2\alpha I_k + I_{k+1} = \int_0^\infty x^{k-1}(\alpha + x)^2 e^{-x^2} \, dx > 0$$

for all real α. Letting $\alpha = -I_k / I_{k+1}$, conclude that

$$I_{2n}^2 < I_{2n-1} \cdot I_{2n+1} \qquad \text{and}$$
$$I_{2n+1}^2 < I_{2n} \cdot I_{2n+2} = \frac{2n+1}{2} I_{2n}^2.$$

Thus

$$\frac{2}{2n+1} I_{2n+1}^2 < I_{2n}^2 < I_{2n-1} \cdot I_{2n+1}.$$

The argument can be completed by using Stirling's and Wallis's formulae. See Young (1992).

2.2 Cauchy's Integral

What exactly did Cauchy accomplish regarding integration? First, he defined a constructive process (an algorithm, if you will) for calculating integrals, and second, he investigated requirements on the function to be integrated that would make his algorithm meaningful. Here is Cauchy's 1823 definition of the integral.

Given a bounded function f on the interval $[a, b]$, divide $[a, b]$ into a finite number of contiguous subintervals $[x_{k-1}, x_k]$ with $a < x_0 < x_1 < \cdots < x_n = b$. We shall use the following terminology.

- The collection of point intervals $(x_0, [x_0, x_1])$, $(x_1, [x_1, x_2]), \ldots,$ $(x_{n-1}, [x_{n-1}, x_n])$ is called a *Cauchy partition* of $[a, b]$, denoted by P.
- We call the points $x_0, x_1, \ldots, x_n$ the *division points* of P.

- The points $x_0, x_1, \ldots, x_{n-1}$ are called the *tags* of P.

Form the Cauchy sum, that is, f evaluated at the tags,

$$\sum_{k=1}^{n} f(x_{k-1})(x_k - x_{k-1}), \quad \sum_{P} f \Delta x.$$

The limit (provided it exists) of such sums, as the lengths of the subintervals approach 0, is said to be the *Cauchy integral of f over* $[a, b]$, written $C \int_a^b f(x)\, dx$. That is, a bounded function f on the interval $[a, b]$ is said to be *Cauchy integrable on* $[a, b]$ iff there is a number A with the property that for each $\epsilon > 0$ there exists a positive constant δ such that, for any Cauchy partition P of $[a, b]$ whose subintervals have length less than δ,

$$\left| \sum_{P} f(x_{k-1})(x_k - x_{k-1}) - A \right| < \epsilon.$$

We write $C \int_a^b f(x)\, dx = A$.

2.2.1 Cauchy's Theorem (1823)

This definition of the Cauchy integral raises an obvious question: Is boundedness of f sufficient to guarantee the existence of A? In a first-of-its-kind theorem, Cauchy argued that continuity of the function would guarantee a successful outcome to this complicated limiting process, a number A.

Theorem 2.2.1 (Cauchy, 1823). *If f is continuous on the interval $[a, b]$, then f is Cauchy integrable on $[a, b]$.*

Proof. The argument hinges on two ideas:

1. Continuous functions on closed bounded intervals are uniformly continuous.

2. Cauchy sequences are useful when we do not have a limit in hand, in this case, the number A in the definition for Cauchy integrability.

From these ideas, we will show that the Cauchy sums, $\sum_P f \Delta x$, converge to a limit A, and that this limit does not depend on the particular choice of partitions, except for the requirement that the lengths of the subintervals of the partitions approach 0.

We will compare Cauchy sums of any two partitions.

Begin with an easier problem: Compare the Cauchy sums of a partition P and another partition $\hat{P}$ obtained by adding a finite number of additional division points to P, a so-called *refinement* of P.

Suppose $a = x_0 < x_1 < \cdots < x_n = b$ are the division points of P with

$$x_0 = y_{10}, y_{11}, y_{12}, \ldots, y_{1i_1} = x_1,$$
$$x_1 = y_{20}, y_{21}, y_{22}, \ldots, y_{2i_2} = x_2,$$
$$\vdots$$
$$x_{n-1} = y_{n0}, y_{n1}, y_{n2}, \ldots, y_{ni_n} = x_n, \qquad \text{the division points of } \hat{P}.$$

Then,

$$\left| \sum_P f\Delta x - \sum_{\hat{P}} f\Delta y \right|$$

$$= \left| \sum_{k=1}^{n} f(x_{k-1})(x_k - x_{k-1}) - \sum_{k=1}^{n} \sum_{j=1}^{i_k} f(y_{kj-1})(y_{kj} - y_{kj-1}) \right|$$

$$= \left| \sum_{k=1}^{n} \sum_{j=1}^{i_k} [f(x_{k-1}) - f(y_{kj-1})](y_{kj} - y_{kj-1}) \right|.$$

Because the function f is uniformly continuous on the interval $[a, b]$, given an $\epsilon > 0$ we have a positive number δ so that $|f(x) - f(y)| < \epsilon$ when x and y are any points of the interval $[a, b]$ within δ of each other. By requiring that the subintervals of P have length less than this δ, we may now conclude that

$$\left| \sum_P f\Delta x - \sum_{\hat{P}} f\Delta y \right| < \epsilon(b - a).$$

Cauchy sums for refinements are within $\epsilon(b-a)$ of each other whenever the subintervals of the partition all have length less than the δ determined by uniform continuity.

Back to the original problem. Suppose that P_1 and P_2 are any partitions of the interval $[a, b]$ whose subintervals have length less than δ. Together, $P_1 \cup P_2$, we have a refinement of each, whose subintervals have length less than δ. Thus the associated sums, $\sum_{P_1} f\Delta x$ and $\sum_{P_2} f\Delta x$, are within $\epsilon(b - a)$ of $\sum_{P_1 \cup P_2} f\Delta x$ and hence within $2\epsilon(b - a)$ of each other.

Any two Cauchy sums differ arbitrarily little from each other iff the lengths of the associated partitions' subintervals are sufficiently small. So take any sequence of partitions P_n of the interval $[a, b]$, with the length of the longest subintervals approaching 0 as $n \to 0$. The sequence of Cauchy sums, $\{\sum_{P_n} f \Delta_n x\}$, are all within $2\epsilon(b - a)$ of each other, for n sufficiently large: $\sum_{P_n} f \Delta_n x \to A$. Show that this limit does not depend on the particular choice of partitions as long as the requirement of the length of the largest subinterval approaching 0 is met.

We have shown that the Cauchy process for continuous functions is well defined.   □

2.2.2   Cauchy Criteria for Cauchy Integrability

In the preceding argument, we actually have the following result.

Theorem 2.2.2. *A continuous function f on the interval $[a, b]$ is Cauchy integrable iff for each $\epsilon > 0$ there exists a positive constant δ so that, if P_1 and P_2 are any Cauchy partitions of $[a, b]$ whose subintervals have length less than δ, the associated Cauchy sums are within ϵ of each other:*

$$\left| \sum_{P_1} f \Delta x - \sum_{P_2} f \Delta y \right| < \epsilon.$$

If f is continuous on $[a, b]$, then $|f|$ is continuous on $[a, b]$, so Cauchy integrability of f implies Cauchy integrability of $|f|$. The converse is false: If

$$f(x) = \begin{cases} 1 & x \text{ rational,} \\ -1 & x \text{ irrational,} \end{cases}$$

then $|f|$ is continuous and $C \int_0^1 |f(x)| \, dx = 1$.

2.3   Recovering Functions by Integration

Now that we have an integration process for continuous functions, is there some means of calculating the Cauchy integral that does not require the genius of Fermat, Stieltjes, et al.?

In many cases the answer is yes: The first part of the so-called Fundamental Theorem of Calculus (FTC) recovers a function from its derivative by integration.

Theorem 2.3.1 (FTC for the Cauchy Integral). *If F is a differentiable function on the interval $[a, b]$, and F' is continuous on $[a, b]$, then*

1. F' is Cauchy integrable on $[a, b]$, and

2. $C \int_a^x F'(t)dt = F(x) - F(a)$ for each x in the interval $[a, b]$.

Proof. The first conclusion follows from Theorem 2.2.1.

To show the second conclusion involves the uniform continuity of F and the mean value theorem for derivatives. Suppose $F(x_k) - F(x_{k-1}) = F'(c_k)(x_k - x_{k-1})$, for $x_{k-1} < c_k < x_k$. Let $\epsilon > 0$ be given. From the Cauchy integrability of F', we have a positive number δ_1 so that if P is any Cauchy partition of the interval $[a, x]$ with subintervals of length less than δ_1, then

$$\left| \sum_P F' \Delta x - C \int_a^x F'(t)dt \right| < \epsilon.$$

Because the derivative F' is continuous by assumption, and thus uniformly continuous on the interval $[a, b]$, we have a positive number δ_2 so that $|F'(c) - F'(d)| < \epsilon$ whenever c and d are points of the interval $[a, b]$ satisfying $|c - d| < \delta_2$.

Letting δ be the smaller of the two numbers δ_1 and δ_2, we may conclude that

$$\left| F(x) - F(a) - C \int_a^x F'(t)dt \right|$$

$$= \left| \sum_{k=1}^n [F(x_k) - F(x_{k-1})] - \sum_{k=1}^n F'(x_{k-1})(x_k - x_{k-1}) \right.$$

$$\left. + \sum_{k=1}^n F'(x_{k-1})(x_k - x_{k-1}) - C \int_a^x F'(t)dt \right|$$

$$\leq \sum_{k=1}^n |F'(c_k) - F'(x_{k-1})| (x_k - x_{k-1}) + \left| \sum_P F' \Delta x - C \int_a^x F'(t)dt \right|$$

$$< \epsilon(b - a) + \epsilon,$$

for all partitions P with subintervals of length less than δ. The proof is complete. □

Compare Theorems 3.7.1, 6.4.2, and 8.7.3.

Exercise 2.3.1. Redo Exercise 2.1.1 in light of this result, and conclude that

$$C \int_0^b x^2 \, dx = \frac{b^3}{3} \quad \text{and}$$

$$C \int_0^\pi \sin x \, dx = \cos 0 - \cos \pi = 2.$$

Exercise 2.3.2. Let

$$F(x) = \begin{cases} x^3 \sin(\pi/x) & 0 < x \le 1, \\ 0 & x = 0. \end{cases}$$

  a. Calculate F'.

  b. Show that F' is continuous on $[0, 1]$.

  c. Calculate $C \int_0^1 F'(x) \, dx$.

  Cauchy not only gave us the existence of the integral for a large class of functions (continuous), but also gave us a straightforward means of calculating many integrals.

2.4   Recovering Functions by Differentiation

In addition to the idea of recovering a function from its derivative by integration, we have the notion of recovering a function from its integral by differentiation, the second part of the Fundamental Theorem of Calculus. Let's examine some properties of the Cauchy integral. Is it continuous? Is it differentiable?...

Theorem 2.4.1 (Another FTC for the Cauchy Integral). *If f is a continuous function on the interval $[a, b]$, and we define a function F on $[a, b]$ by $F(x) = C \int_a^x f(t) \, dt$, then*

  1. F is differentiable on $[a, b]$,

  2. $F' = f$ on $[a, b]$, and

  3. F is absolutely continuous on $[a, b]$.

Proof. We have shown that the Cauchy integral of a continuous function is well defined, so F makes sense. To show that F is differentiable on $[a, b]$, and that the derivative of F, F', is in fact equal to f, entails estimating the familiar expression

$$\left| \frac{F(x + h) - F(x)}{h} - f(x) \right|$$

where h is small enough so that $x + h$ belongs to the interval (a, b).

Because f is continuous at x by assumption, we have a $\delta > 0$ so that $|f(t) - f(x)| < \epsilon$ whenever $|t - x| < \delta$ and t belongs to the interval $[a, b]$. Then,

$$\left| \frac{F(x + h) - F(x)}{h} - f(x) \right| = \left| \frac{1}{h} \cdot C \int_x^{x+h} [f(t) - f(x)] \, dt \right| < \epsilon.$$

In other words, $\left(C \int_a^x f(t) \, dt \right)' = f(x)$, and this is what we wanted to show.

As for the absolute continuity, see Definition 5.8.2 and note that F' is continuous and thus bounded on the interval $[a, b]$:

$$|F(b_k) - F(a_k)| = |F'(c_k)(b_k - a_k)| \leq B \, |b_k - a_k|. \quad \square$$

Compare Theorems 3.7.2, 6.4.1, and 8.8.1.

Exercise 2.4.1. Given

$$f(x) = \begin{cases} 1 - x & 0 \leq x \leq 1, \\ 2x - 2 & 1 \leq x \leq 2. \end{cases}$$

a. Determine $F(x) = C \int_0^x f(t) \, dt$, $G(x) = C \int_0^{2x} f(t) \, dt$, $H(x) = C \int_0^{1/2x} f(t) \, dt$.

b. Using the definition of the derivative, show that F is differentiable at $x = 1$ and $F'(1) = f(1)$.

c. Determine G' and H'.

Integration is a smoothing process: Continuous functions become differentiable functions under the integration process.

2.5   A Convergence Theorem

Another means of calculating integrals is using convergence theorems, comparing the integrals of a sequence of functions with the integral of the limit of the sequence of functions. This is valid with some restrictions.

Theorem 2.5.1 (Convergence for Cauchy Integrable Functions).   *If $\{f_k\}$ is a sequence of continuous functions converging uniformly to the function f on $[a, b]$, then f is Cauchy integrable on $[a, b]$ and $C \int_a^b f(x) \, dx = \lim C \int_a^b f_k(x) \, dx$.*

Proof. That the function f is continuous follows from Weierstrass's Theorem: The uniform limit of a sequence of continuous functions is continuous:

$$|f(x) - f(y)| \leq |f(x) - f_k(x)| + |f_k(x) - f_k(y)| + |f_k(y) - f(y)|,$$

the first and third terms are "small" by uniform convergence, and the second term is "small" by continuity of f_k. Thus f is Cauchy integrable on $[a, b]$.

As for the second conclusion,

$$\left| C \int_a^b f_k(x)\, dx - C \int_a^b f(x)\, dx \right| \leq C \int_a^b |f_k(x) - f(x)|\, dx,$$

and the right-hand side of this inequality can be made arbitrarily small by uniform convergence: Given $\epsilon > 0$, we have a natural number K so that $|f_k - f| < \epsilon$ whenever $k \geq K$, throughout the interval $[a, b]$. $\square$

Exercise 2.5.1. Consider a sequence of continuous functions $\{f_k\}$ given by

$$f_1(x) = \begin{cases} 16x & 0 \leq x \leq \frac{1}{4}, \\ 8 - 16x & \frac{1}{4} \leq x \leq \frac{1}{2}, \\ 0 & \frac{1}{2} \leq x \leq 1; \end{cases} \qquad f_2(x) = \begin{cases} f_1(x) & 0 \leq x \leq \frac{1}{2}, \\ 8 - 16x & \frac{1}{2} \leq x \leq \frac{5}{8}, \\ 16x - 12 & \frac{5}{8} \leq x \leq \frac{3}{4}, \\ 0 & \frac{3}{4} \leq x \leq 1; \end{cases}$$

$$f_3(x) = \begin{cases} f_2(x) & 0 \leq x \leq \frac{3}{4}, \\ 16x - 12 & \frac{3}{4} \leq x \leq \frac{13}{16}, \\ 14 - 16x & \frac{13}{16} \leq x \leq \frac{7}{8}, \\ 0 & \frac{7}{8} \leq x \leq 1; \end{cases} \qquad \dots .$$

a. Graph f_1, f_2, and f_3.

b. Show that the sequence $\{f_k\}$ converges uniformly on $[0, 1]$, where $|f_k - f_{k+1}| \leq 2^{2-k}$.

c. Calculate $C \int_0^1 \lim f_k(x)\, dx$.

Exercise 2.5.2. Let the function $f_k(x) = x^{1/k}$, where $k = 1, 2, \dots$ and $0 \leq x \leq 1$.

a. Calculate $\lim f_k = f$. Is the convergence uniform?

b. Calculate $\lim C \int_0^1 f_k(x)\, dx$.

2.6   Joseph Fourier

We now go back to the year 1822, in which year the book *Théorie ana-lytique de la chaleur*, by Joseph Fourier (1768–1830), appeared. Fourier was looking for a steady-state temperature function $T(x, y)$ in the strip, $0 \leq x \leq \pi, 0 \leq y$, satisfying the partial differential equation,

$$\frac{\partial^2 T}{\partial x^2} + \frac{\partial^2 T}{\partial y^2} = 0 \qquad \text{for } 0 < x < \pi, \, 0 < y,$$

with temperature 0 when $x = 0$ or $x = \pi$, and an initial temperature distribution ϕ on the base where $y = 0$.

Example 2.6.1. Assuming that $T(x, y) = X(x)Y(y)$ (separation of variables), we argue that $T_k(x) = X_k(x)Y_k(y) = e^{-ky} \sin(kx)$, for $k = 1, 2,$ ..., satisfies the partial differential equation and that $T_k(0) = T_k(\pi) = 0$.

Heuristically, $T(x, y) = \sum b_k e^{-ky} \sin(kx)$, the constants b_k to be determined. To satisfy the base condition at $y = 0$ we need $\phi(x) = \sum b_k \sin(kx)$ for $0 < x < \pi$. Multiply both sides of the equality by $\sin(nx)$. Assuming that the integrals are meaningful and that we can integrate termwise, show that the coefficients b_n are given by $b_n = 2/\pi \int_0^\pi \phi(x) \sin(nx) \, dx$, for $0 < x < \pi$.

What kind of integral are we discussing here? Certainly if ϕ is continuous, the Cauchy integral would be sufficient. But periodicity and sine suggest an odd extension of ϕ. If $\phi(0) \neq 0$, we have problems at $x = 0$.

Because Fourier series successfully model a wide variety of physical phenomena, we want to be able to deal with discontinuous behavior.

Exercise 2.6.1. a. Suppose

$$f(x) = \begin{cases} 1 & 0 \leq x < 1 \text{ or } 1 < x \leq 2, \\ 2 & x = 1. \end{cases}$$

Because f is discontinuous, f is not Cauchy integrable. On the other hand, show that for any Cauchy partition P of $[0, 2]$,

$$\left| \sum_P f(x_{k-1}) \Delta x - 2 \right| \leq \delta.$$

b. Suppose

$$f(x) = \begin{cases} 1 & 0 \leq x < 1, \\ 2 & 1 \leq x \leq 2. \end{cases}$$

Show that for any Cauchy partition of $[0, 2]$, $\left| \sum f \Delta x - 3 \right| \leq \delta$.

Argue that for a bounded function with a finite number of jump discontinuities, obvious modifications to the Cauchy integral could be made.

2.7   P. G. Lejeune Dirichlet

Several mathematicians tried to discover conditions that would guarantee convergence of the Fourier series. A successful argument was given by P.G. Lejeune Dirichlet (1805–1859).

Theorem 2.7.1 (Dirichlet's Convergence Theorem for Fourier Series, 1829). *Suppose that f is a bounded function, piecewise continuous (finite number of jump discontinuities) and piecewise monotonic on the interval $[-\pi, \pi]$, with period 2π. The Fourier series representation of f at x,*

$$\frac{1}{2}a_0 + \sum_{1}^{\infty}(a_n \cos nx + b_n \sin nx)$$

with

$$a_n = \frac{1}{\pi}\int_{-\pi}^{\pi} f(x)\cos(nx)\,dx, \qquad n = 0, 1, 2, \ldots, \text{ and}$$

$$b_n = \frac{1}{\pi}\int_{-\pi}^{\pi} f(x)\sin(nx)\,dx, \qquad n = 1, 2, \ldots,$$

converges to

$$\frac{f(x+0) + f(x-0)}{2},$$

where, as usual, $\lim_{y\to x+} f(y) = f(x+0)$ *and* $\lim_{y\to x-} f(y) = f(x-0)$.

For a discussion and proof, see Bressoud (1994).

Exercise 2.7.1. Assuming Dirichlet's result, with $f(x) = |x|$ for $-\pi \le x \le \pi$, extended periodically, show the following:

$$|x| = \frac{\pi}{2} - \frac{4}{\pi}[\cos(x) + \frac{\cos(3x)}{3^2} + \frac{\cos(5x)}{5^2} + \cdots] \qquad for \ -\pi \le x \le \pi,$$

$$\frac{\pi^2}{8} = 1 + \frac{1}{3^2} + \frac{1}{5^2} + \cdots,$$

$$\frac{\pi^2}{6} = 1 + \frac{1}{2^2} + \frac{1}{3^2} + \cdots. \qquad \text{(Euler)}$$

Exercise 2.7.2. Suppose $f(x) = \cos(\alpha x)$, for $-\pi \le x \le \pi$ and α not an integer, extended periodically. (From Courant and John (1965).)

a. Show that

$$\cos(\alpha x) = \frac{2\alpha \sin(\alpha x)}{\pi} \left(\frac{1}{2\alpha^2} - \frac{\cos(x)}{\alpha^2 - 1^2} + \frac{\cos(2x)}{\alpha^2 - 2^2} - \cdots \right).$$

b. Let $x = \pi$ and conclude

$$\cot(\pi x) - \frac{1}{\pi x} = \frac{-2x}{\pi} \left(\frac{1}{1^2 - x^2} + \frac{1}{2^2 - x^2} + \cdots \right).$$

c. For $0 < x \le \beta < 1$, show that the series converges uniformly. Then integrate term by term to show

$$\ln \left(\frac{\sin(\pi x)}{\pi x} \right) = \lim_n \sum_{k=1}^{n} \ln \left(1 - \frac{x^2}{k^2} \right).$$

So

$$\frac{\sin(\pi x)}{\pi x} = \left(1 - \frac{x^2}{1^2} \right) \left(1 - \frac{x^2}{2^2} \right) \left(1 - \frac{x^2}{3^2} \right) \cdots, \qquad \text{for } 0 < x < 1.$$

d. Show that

$$\frac{\pi}{2} = \left(\frac{2}{1} \right) \left(\frac{2}{3} \right) \left(\frac{4}{3} \right) \left(\frac{4}{5} \right) \cdots. \qquad \text{(Wallis)}$$

While studying convergence problems for Fourier series, Dirichlet began to consider functions that assumed one value on the rationals and a different value on the irrationals. For example, suppose

$$f(x) = \begin{cases} 1 & x \text{ rational,} \\ 0 & x \text{ irrational,} \end{cases} \qquad \text{and} \qquad g(x) = \begin{cases} 0 & x \text{ rational,} \\ 1 & x \text{ irrational,} \end{cases}$$

Certainly $1 = C \int_0^1 (f + g)(x) \, dx$, and linearity of the integral would require that $1 = C \int_0^1 f(x) \, dx + C \int_0^1 g(x) \, dx$. But these two Cauchy integrals are not defined.

Dirichlet had discussions with Riemann to try to find an integration process that would overcome that difficulty. Riemann did not find such an integration process (that would be discovered by Lebesgue), but did develop another integration process more powerful than Cauchy's. Riemann's process is the subject of the next chapter.

2.8   Patrick Billingsley's Example

Lest we become blasé about results concerning continuous functions, here is Patrick Billingsley's (1982) *continuous nowhere differentiable function.*

   Suppose $\phi(x)$ is the distance from x to the nearest integer. Define a function B by $B(x) = \sum_{n=0}^{\infty} \phi(2^n x)/2^n$. That B is continuous follows from the Weierstrass M-test.

Exercise 2.8.1. Calculate $C \int_0^1 B(x)\, dx$.

   In Exercise 5.9.2 we shall see that if a function f has a derivative at x_0, then for every $\epsilon > 0$ we may determine a positive number δ so that

$$\left| f(\beta) - f(\alpha) - f'(x_0)(\beta - \alpha) \right| < \epsilon(\beta - \alpha),$$

if $\alpha \neq \beta$ and $x_0 - \delta < \alpha \leq x_0 \leq \beta < x_0 + \delta$.

   To show B is not differentiable, we will pick a point x_0 in the interval $[0, 1)$ and approximate it with binary expansions. Because x_0 is between 0 and $\frac{1}{2}$ or $\frac{1}{2}$ and 1, we have

$$\frac{a_1}{2} \leq x_0 < \frac{a_1}{2} + \frac{1}{2}, \qquad \text{for } a_1 = 0 \text{ or } 1.$$

Now divide $[0, 1]$ into four subintervals of equal length $1/2^2$. Then

$$\frac{a_1}{2} + \frac{a_2}{2^2} \leq x_0 < \frac{a_1}{2} + \frac{a_2}{2^2} + \frac{1}{2^2}, \qquad \text{for } a_1, a_2 = 0 \text{ or } 1.$$

Continuing,

$$\frac{a_1}{2} + \cdots + \frac{a_N}{2^N} \leq x_0 < \frac{a_1}{2} + \cdots + \frac{a_N}{2^N} + \frac{1}{2^N} \qquad \text{for } a_i = 0 \text{ or } 1.$$

   Let

$$\alpha_N = \frac{a_1}{2} + \cdots + \frac{a_N}{2^N} \leq x_0 < \frac{a_1}{2} + \cdots + \frac{a_N}{2^N} + \frac{1}{2^N} = \beta_N,$$

and form the difference quotient

$$\frac{B(\beta_N) - B(\alpha_N)}{\beta_N - \alpha_N} = \sum_{n=0}^{N-1} \frac{\phi(2^n \beta_N) - \phi(2^n \alpha_N)}{2^n \beta_N - 2^n \alpha_N} + \sum_{n=N}^{\infty} \frac{\phi(2^n \beta_n) - \phi(2^n \alpha_N)}{2^n(\beta_N - \alpha_N)}.$$

   We have $2^n \beta_N = (m+1)/2^{N-n}$ and $2^n \alpha_N = m/2^{N-n}$, for m an integer. Also, ϕ is linear with slope ± 1 on the interval $[m/2^{N-n}, (m+1)/2^{N-n}]$

as long as $N - n \geq 1$, that is, when $0 \leq n \leq N - 1$. Thus the first sum is a sum of ± 1s:

$$\sum_{n=0}^{N-1} \frac{\phi(2^n \beta_N) - \phi(2^n \alpha_N)}{2^n \beta_N - 2^n \alpha_N} = \sum_{n=0}^{N-1} \pm 1.$$

The second sum vanishes because $2^n \beta_N = (m + 1)2^{n-N}$ and $2^n \alpha_N = m2^{n-N}$ are integers for $n \geq N$.

Letting $N \to \infty$, $\beta_N - \alpha_N \to 0$, differentiability of B at x_0 would yield $B'(x_0) = \sum \pm 1$.

2.9   Summary

Two Fundamental Theorems of Calculus for the Cauchy Integral:
   If F is differentiable on $[a, b]$ and if F' is continuous on $[a, b]$, then

1. F' is Cauchy integrable on $[a, b]$ and

2. $C \int_a^x F'(t)dt = F(x) - F(a)$, for $a \leq x \leq b$.

If f is continuous on $[a, b]$ and $F(x) = C \int_a^x f(t)dt$, then

1. F is differentiable on $[a, b]$,

2. $F' = f$ on $[a, b]$, and

3. F is absolutely continuous on $[a, b]$.

2.10   References

1. Billingsley, Patrick. Van der Waerden's continuous nowhere differentiable function. *American Mathematical Monthly* 89 (1982) 691.

2. Bressoud, David. *A Radical Approach to Real Analysis*. Washington: Mathematical Association of America, 1994.

3. Courant, Richard, and Fritz John. *Introduction to Calculus and Analysis*. Vol. 1. New York: Wiley Interscience, 1965.

4. Young, Robert M. *Excursions in Calculus: An Interplay of the Continuous and the Discrete*. Dolciani Mathematical Expositions, No. 13. Washington: Mathematical Association of America, 1992.

CHAPTER 3

The Riemann Integral

Reason with a capital R = Sweet Reason, the newest and rarest thing in human life, the most delicate child of human history.

— Edward Abbey

We are nature's unique experiment to make the rational intelligence prove itself sounder than the reflex. Knowledge is our destiny.

— Jacob Bronowski

The Riemann integral (1854) — the familiar integral of calculus developed by Bernhard Riemann (1826–1866) — was a response to various questions raised by Dirichlet about just how discontinuous a function could be and still have a well-defined integral.

3.1 Riemann's Integral

Given a bounded function f on the interval $[a, b]$, divide $[a, b]$ into a finite number of contiguous subintervals $[x_{k-1}, x_k]$, with $a = x_0 < x_1 < \cdots < x_n = b$, and pick a point c_k in $[x_{k-1}, x_k]$. As with Cauchy integration, we begin by defining some terminology.

The collection of point intervals

$$(c_1, [x_0, x_1]),\ (c_2, [x_1, x_2]), \ldots, (c_n, [x_{n-1}, x_n])$$

is called a *Riemann partition* of $[a, b]$, to be denoted by P, the points $x_0, x_1, \ldots, x_n$ are called the *division points* of P, and the points $c_1, c_2, \ldots, c_n$ are called the *tags* of P.

Form the Riemann sum $\sum_{k=1}^{n} f(c_k)(x_k - x_{k-1})$, which we will write as $\sum_P f \, \Delta x$. The limit (provided it exists) of such sums, as the lengths

45

of the subintervals approach zero, is said to be the *Riemann integral of f over* $[a, b]$, written $R \int_a^b f(x) \, dx$.

That is, a bounded function f on the interval $[a, b]$ is said to be *Riemann integrable* on $[a, b]$ if there is a number A with the property that for each $\epsilon > 0$ there exists a positive constant δ such that for any Riemann partition P of $[a, b]$ whose subintervals have length less than δ,

$$\left| \sum_P f \, \Delta x - A \right| < \epsilon.$$

We write $R \int_a^b f(x) \, dx = A$.

Dirichlet-type functions — say, 1 on the irrationals and 0 on the rationals — show again that some restrictions must be imposed on f besides boundedness. In comparing Riemann's and Cauchy's integration processes, we notice that Riemann's f may be evaluated at any point c_k in the interval $[x_{k-1}, x_k]$, whereas Cauchy's f is evaluated at the left-hand endpoint of that interval, x_{k-1}.

For a given partition P, how much variability can we have in the associated Riemann sums as c_k varies throughout the interval $[x_{k-1}, x_k]$? Certainly

$$\sum_P \inf_{[x_{k-1}, x_k]} f \, \Delta x \le \sum_P f(c) \, \Delta x \le \sum_P \sup_{[x_{k-1}, x_k]} f \, \Delta x.$$

Thus the absolute value of the difference of any two Riemann sums for the same partition is bounded above by $\sum_P (\sup f - \inf f) \, \Delta x$, where the quantity $\sup f - \inf f$ denotes $\sup_{[x_{k-1}, x_k]} f - \inf_{[x_{k-1}, x_k]} f$.

On the other hand, we have points c and d in the interval $[x_{k-1}, x_k]$ so that

$$\sup f - \inf f - 2\epsilon < f(c_k) - f(d_k) \le \sup f - \inf f$$

for ϵ an arbitrary positive number. Then

$$\sum_P (\sup f - \inf f) \, \Delta x - 2\epsilon(b - a) < \sum_P f(c) \, \Delta x - \sum_P f(d) \, \Delta x$$

$$\le \sum_P (\sup f - \inf f) \, \Delta x.$$

Thus $\sum_P (\sup f - \inf f) \, \Delta x$ is the smallest upper bound of the absolute value of the difference of any two Riemann sums associated with the partition P. (The reader may want to review the Cauchy integral discussion in Section 2.2.)

Proceeding in the same manner, but omitting most of the details, we again have that for any refinement $\hat{P}$ of P,

$$\left| \sum_{k=1}^{n} f(c_k)\,\Delta x - \sum_{k=1}^{n}\sum_{j=1}^{k} f(c_{kj})\,\Delta y \right|$$

$$= \left| \sum_{k=1}^{n}\sum_{j=1}^{i_k} [f(c_k) - f(c_{kj})]\,\Delta y \right|$$

$$\leq \sum_{k=1}^{n}\sum_{j=1}^{i_k} \left(\sup_{[x_{k-1},x_k]} f - \inf_{[x_{k-1},x_k]} f \right)\,\Delta y$$

$$= \sum_{k=1}^{n} (\sup f - \inf f)\,\Delta x$$

$$= \sum_{P} (\sup f - \inf f)\,\Delta x.$$

That is, for a partition P of $[a,b]$ and any of its refinements, the absolute value of the difference of any two associated Riemann sums has a smallest upper bound of $\sum_P (\sup f - \inf f)\,\Delta x$. Apparently we have an integrability condition.

3.2 Criteria for Riemann Integrability

Let us examine the criteria for what makes a function Riemann integrable.

Theorem 3.2.1 (Riemann Integrability Criteria). *A bounded function f on the interval $[a,b]$ is Riemann integrable iff given an $\epsilon > 0$ we may determine a positive number δ so that $\sum_P (\sup f - \inf f)\,\Delta x < \epsilon$ for all partitions P whose subintervals have length less than δ.*

Proof. If we assume that f is Riemann integrable on the interval $[a,b]$, then given an $\epsilon > 0$ we have a $\delta > 0$ and a number A so that for any partition of $[a,b]$ whose subintervals have length less than δ, every Riemann sum is between $A - \epsilon/4$ and $A + \epsilon/4$. But then the absolute value of the difference of any two Riemann sums is less than $\epsilon/2$, and $\sum_P (\sup f - \inf f)\,\Delta x < \epsilon$.

On the other hand, suppose that given an $\epsilon > 0$ we have a $\delta > 0$ so that $\sum_P (\sup f - \inf f)\,\Delta x < \epsilon/2$ for all partitions P whose subintervals have lengths less than δ. Let P_1 and P_2 be two such partitions:

$$\sum_{P_1} (\sup f - \inf f)\,\Delta x < \frac{\epsilon}{2} \quad \text{and} \quad \sum_{P_2} (\sup f - \inf f)\,\Delta y < \frac{\epsilon}{2}.$$

Since $P_1 \cup P_2$ is a refinement of P_1 and P_2, all of its subintervals will have lengths less than δ. Furthermore,

$$\left| \sum_{P_1} f\, \Delta x - \sum_{P_2} f\, \Delta y \right|$$

$$\leq \left| \sum_{P_1} f\, \Delta x - \sum_{P_1 \cup P_2} f\, \Delta z \right| + \left| \sum_{P_1 \cup P_2} f\, \Delta z - \sum_{P_2} f\, \Delta y \right|$$

$$\leq \sum_{P_1} (\sup f - \inf f)\, \Delta x + \sum_{P_2} (\sup f - \inf f)\, \Delta y < \epsilon.$$

Associated Riemann sums for partitions whose subintervals have lengths less than δ are "close." Consider any sequence of partitions P_n of the interval $[a, b]$, with δ_n approaching 0 as $n \to \infty$, and form a sequence of associated Riemann sums: $\{\sum_{P_n} f\, \Delta_n x\}$. For n sufficiently large, $\delta_n < \delta$, and consequently

$$\left| \sum_{P_n} f\, \Delta_n x - \sum_{P_m} f\, \Delta_m x \right| < \epsilon.$$

The sequence $\{\sum_{P_n} f\, \Delta_n x\}$ is a Cauchy sequence. We have that

$$\sum_{P_n} f\, \Delta_n x \to A.$$

Show that another sequence $\hat{P}_n$ of partitions with $\hat{\delta}_n$ approaching 0 as $n \to \infty$ leads to the same result. This completes the argument. □

Exercise 3.2.1. Suppose that $f(x) = x^2$, $0 \leq x \leq 1$. Show that f is Riemann integrable on $[0, 1]$. Hint: $\sum_P \left(x_k^2 - x_{k-1}^2 \right) \Delta x < \sum_P 2x_k \Delta x \Delta x < 2\delta$.

Exercise 3.2.2. If f is continuous on $[a, b]$, then f is Riemann integrable on $[a, b]$. Also, if f is Cauchy integrable on $[a, b]$, then f is Riemann integrable on $[a, b]$. Hint: Let $\epsilon > 0$ be given. Because f is uniformly continuous on $[a, b]$, we have a $\delta > 0$, so that $|f(c) - f(d)| < \epsilon$ whenever $|c - d| < \delta$. Let P be any partition of $[a, b]$ whose subintervals have length less than δ.

Exercise 3.2.3. Cauchy integrable functions are Riemann integrable functions. Do the integrals have the same value? Hint:

$$
C \int_a^b f(x)\,dx - R \int_a^b f(x)\,dx = \left[C \int_a^b f(x)\,dx - \sum f(x_{k-1})\,\Delta x \right]
$$
$$
+ \left[\sum f(x_{k-1})\,\Delta x - \sum f(c_k)\,\Delta x \right]
$$
$$
+ \left[\sum f(c_k)\,\Delta x - R \int_a^b f(x)\,dx \right],
$$

and use the uniform continuity of f on $[a, b]$.

3.3 Cauchy and Darboux Criteria for Riemann Integrability

Just how discontinuous can a bounded function be and maintain Riemann integrability? We have more work to do before we can answer this question. First, we will somewhat improve the Riemann Integrability Criteria.

Theorem 3.3.1 (Cauchy Criteria for Riemann Integrability). *A bounded function f on the interval $[a, b]$ is Riemann integrable iff for each $\epsilon > 0$ there exists a Riemann partition P of $[a, b]$ so that $\sum_P (\sup f - \inf f)\,\Delta x < \epsilon$.*

Proof. Let $\epsilon > 0$ be given, with P^* a Riemann partition of $[a, b]$, where $a = x_0^* < x_1^* < \cdots < x_M^* = b$, so that $\sum_{P*} (\sup f - \inf f)\,\Delta x < \epsilon$. We will construct a positive constant δ so that any partition P of $[a, b]$ whose subintervals have length less than δ satisfies $\sum_P (\sup f - \inf f)\,\Delta x < \epsilon$.

Suppose P has division points $a = x_0 < x_1 < \cdots < x_n = b$. Form the refinement $P \cup P^*$ of P. Initially, suppose this refinement has one more division point than P, say $x_{k-1} < x_J^* < x_k$ for some k. In this case,

$$
\sum_{P \cup P^*} \inf f\,\Delta z - \sum_P \inf f\,\Delta x = \left(\inf_{[x_{k-1}, x_J^*]} f - \inf_{[x_{k-1}, x_k]} f \right) \left(x_J^* - x_{k-1} \right)
$$
$$
+ \left(\inf_{[x_J^*, x_k]} f - \inf_{[x_{k-1}, x_k]} f \right) \left(x_k - x_J^* \right)
$$
$$
\le 2B\,(x_k - x_{k-1}),
$$

where $|f| < B$ on the interval $[a, b]$.

Now, since P^* consists of $M + 1$ division points, we can add at most $M - 1$ such points to P different from a and b. Thus for the partition P,

$$\sum_{P \cup P^*} \inf f \, \Delta z - \sum_{P} \inf f \, \Delta x$$
$$< (M - 1)2B \cdot \text{(maximum length of the subintervals of } P).$$

Since $\sum_{P^*} \inf f \, \Delta x^* \leq \sum_{P \cup P^*} \inf f \, \Delta z$, we have

$$\sum_{P^*} \inf f \, \Delta x^* - \sum_{P} \inf f \, \Delta x$$
$$< 2B(M - 1) \cdot \text{(maximum length of the subintervals of } P).$$

The reader may show

$$\sum_{P} \sup f \, \Delta x - \sum_{P^*} \sup f \, \Delta x^*$$
$$< 2B(M - 1) \cdot \text{(maximum length of the subintervals of } P).$$

Adding, we have

$$\sum_{P} (\sup f - \inf f) \, \Delta x < \sum_{P^*} (\sup f - \inf f) \, \Delta x^* + 4B(M - 1)$$
$$\cdot \text{(maximum length of the subintervals of } P)$$
$$< \epsilon + \epsilon,$$

provided we choose δ so that $4B(M - 1)\delta < \epsilon$. This is not a problem, since we began with $|f| < B$ and a partition P^* with $M + 1$ division points. The proof in the other direction follows from Theorem 3.2.1. □

Exercise 3.3.1. Use the preceding result to show that $R \int_0^1 x^2 \, dx = \frac{1}{3}$. Hint: Consider $0, 1/n, 2/n, \ldots, n/n$.

Theorem 3.3.1 suggests another approach to the Riemann integral (Darboux, 1875). For a function f bounded on $[a, b]$ and any partition P of $[a, b]$,

$$\inf_{[a,b]} f (b - a) \leq \sum \inf f \, \Delta x \leq \sum \sup f \, \Delta x \leq \sup_{[a,b]} f (b - a).$$

The lower (upper) sum of f with respect to the partition P is defined as

$$L(f, P) \equiv \sum \inf f \, \Delta x,$$
$$(U(f, P) \equiv \sum \sup f \, \Delta x).$$

The lower sums are bounded above by $\sup_{[a,b]} f(b-a)$, and the upper sums are bounded below by $\inf_{[a,b]} f(b-a)$.

Define the lower Darboux integral of f on $[a,b]$ by

$$\underline{\int_a^b} f(x)dx = \sup_P L(f,P)$$

and the upper Darboux integral of f on $[a,b]$ as

$$\overline{\int_a^b} f(x)dx = \inf_P U(f,P).$$

The bounded function f is said to be Darboux integrable on $[a,b]$ if the numbers $\underline{\int_a^b} f(x)dx, \overline{\int_a^b} f(x)dx$ are equal. In this case we write $D\int_a^b f(x)dx$.

Exercise 3.3.2. Show the following.

  a. For any partition P of $[a,b]$, $L(f,P) \leq U(f,P)$.

  b. Refinements do not decrease lower sums or increase upper sums: If P^* is a refinement of P, $L(f,P) \leq L(f, P \cup P^*)$ and $U(f, P \cup P^*) \leq U(f,P)$. Hint: $P^* = P \cup \{z\}$, $x_{J-1} \leq z \leq x_J$.

  c. Any lower sum does not exceed any upper sum: $L(f, P_1) \leq U(f, P_2)$. Hint: Part b, $P_1 \cup P_2$.

  d. $\underline{\int_a^b} f(x)dx \leq \overline{\int_a^b} f(x)dx$. Hint: $L(f, P_1) \leq U(f, P_2)$ for any partitions of $[a,b]$. "Fix" P_1 and vary P_2. Thus $L(f, P_1) \leq \overline{\int_a^b} f(x)dx$. Vary P_1, and so on.

Theorem 3.3.2 (Darboux Integrability Criteria). *A bounded function f on the interval $[a,b]$ is Darboux integrable iff for each $\epsilon > 0$ there exists a Riemann partition P_ϵ of $[a,b]$ so that $\sum_{P_\epsilon} (\sup f - \inf f)\Delta x < \epsilon$.*

Proof. Let $\epsilon > 0$ be given. Assume we have a partition P_ϵ of $[a,b]$ so that $\sum_{P_\epsilon} (\sup f - \inf f)\Delta x < \epsilon$; that is, $U(f, P_\epsilon) - L(f, P_\epsilon) < \epsilon$. However,

$$L(f, P_\epsilon) \leq \underline{\int_a^b} f(x)dx \leq \overline{\int_a^b} f(x)dx \leq U(f, P_\epsilon).$$

Thus $\overline{\int_a^b} f(x)dx - \underline{\int_a^b} f(x)dx < \epsilon$. We may conclude

$$\underline{\int_a^b} f(x)dx = \overline{\int_a^b} f(x)dx.$$

We have Darboux integrability.

Now assume we have Darboux integrability:

$$\underline{\int_a^b} f(x)dx = \overline{\int_a^b} f(x)dx = D\int_a^b f(x)dx.$$

Let $\epsilon > 0$ be given. Because

$$\sup L(f,P) = \underline{\int_a^b} f(x)dx = D\int_a^b f(x)dx = \overline{\int_a^b} f(x)dx = \inf U(f,P),$$

we have partitions P_1, P_2 of $[a,b]$ so that

$$D\int_a^b f(x)dx - \frac{\epsilon}{2} < L(f,P_1) \le L(f,P_1 \cup P_2) \le U(f,P_1 \cup P_2)$$

$$\le U(f,P_2) < D\int_a^b f(x)dx + \frac{\epsilon}{2}.$$

We have a partition $P_1 \cup P_2 = P_\epsilon$ of $[a,b]$ with $U(f,P_\epsilon) - L(f,P_\epsilon) < \epsilon$, that is

$$\sum_{P_\epsilon} (\sup f - \inf f)\Delta x < \epsilon. \quad \square$$

Note: Theorems 3.3.1 and 3.3.2 tell us that Riemann integrability is equivalent to Darboux integrability.

3.4   Weakening Continuity

To achieve smallness of the sum $\sum_P (\sup f - \inf f)\,\Delta x$ we have required smallness of $\sup f - \inf f$ on every subinterval of a partition. Is this necessary?

For example, divide a given partition into two collections of subintervals. In the first collection, put the subintervals with "small" $\sup f - \inf f$. This, we hope, will be most of the subintervals in terms of total length close to $b-a$. In the other collection, put the subintervals with "large" $\sup f - \inf f$ (although bounded by $\sup_{[a,b]} f - \inf_{[a,b]} f$). Their total length is, we hope, "small" in comparison with $b - a$. This sorting was Riemann's idea.

Theorem 3.4.1 (Riemann's Theorem). *Let f be a bounded function on the interval $[a,b]$, and let ω and l be any positive numbers. Then f is Riemann integrable on $[a,b]$ iff we have a positive constant δ so that, for* *any partition of $[a, b]$ into subintervals of length less than δ, the sum of the lengths of the subintervals $[x_{k-1}, x_k]$ with $\sup_{[x_{k-1}, x_k]} f - \inf_{[x_{k-1}, x_k]} f$ greater than ω is less than l.*

Proof. We begin by assuming that f is Riemann integrable on $[a, b]$, with ω and l positive numbers. Apply the integrability criteria from Theorem 3.2.1. Then, given an $\epsilon > 0$, we have a $\delta > 0$ so that if P is any partition of $[a, b]$ whose subintervals have lengths less than δ, then $\sum_P (\sup f - \inf f) \Delta x < \epsilon$.

Let $\epsilon = \omega l$. We have a $\hat{\delta}$ for this ϵ so that if P is any partition of $[a, b]$ whose subintervals have lengths less than $\hat{\delta}$,

$$\sum_P (\sup f - \inf f) \Delta x$$

$$= \sum_{\sup f - \inf f \leq \omega} (\sup f - \inf f) \Delta x + \sum_{\sup f - \inf f > \omega} (\sup f - \inf f) \Delta x$$

$$< \omega l.$$

But then

$$\omega \sum_{\sup f - \inf f > \omega} \Delta x < \sum_{\sup f - \inf f > \omega} (\sup f - \inf f) \Delta x < \omega l.$$

That is, the sum of the lengths of the subintervals with $\sup f - \inf f > \omega$ is less than l.

For the other direction, let $\epsilon > 0$ be given, and let

$$\omega = \frac{\epsilon}{2(b - a)}, \qquad l = \frac{\epsilon}{2 \left(\sup_{[a,b]} f - \inf_{[a,b]} f \right)}.$$

The theorem is trivial if f is constant, so we may assume $\sup_{[a,b]} f - \inf_{[a,b]} f > 0$.

By assumption we have a $\delta > 0$ so that for any partition of $[a, b]$ into subintervals whose lengths are less than δ we can describe the sum of the lengths of the subintervals with $\sup f - \inf f$ greater than $\epsilon/[2(b - a)]$:

$$\sum_{\sup f - \inf f > \frac{\epsilon}{2(b-a)}} \Delta x < \frac{\epsilon}{2 \left(\sup_{[a,b]} f - \inf_{[a,b]} f \right)}.$$

But then

$$\sum_P (\sup f - \inf f)\, \Delta x$$

$$= \sum_{\sup f - \inf f \leq \frac{\epsilon}{2(b-a)}} (\sup f - \inf f)\, \Delta x$$

$$+ \sum_{\sup f - \inf f > \frac{\epsilon}{2(b-a)}} (\sup f - \inf f)\, \Delta x$$

$$< \frac{\epsilon}{2(b-a)} \sum \Delta x + \left(\sup_{[a,b]} f - \inf_{[a,b]} f \right) \sum_{\sup f - \inf f > \frac{\epsilon}{2(b-a)}} \Delta x$$

$$< \frac{\epsilon}{2} + \frac{\epsilon}{2}.$$

This concludes the argument. □

Exercise 3.4.1. Given

$$f(x) = \begin{cases} 1 & \dfrac{1}{2n} < x \leq \dfrac{1}{2n-1}, n = 1, 2, \ldots, \\ 0 & \text{otherwise.} \end{cases}$$

a. Use Riemann's Theorem to show that f is Riemann integrable on $[0, 1]$. Hint: Because $\sup f - \inf f \leq 1$ on any subinterval of $[0, 1]$ and $\sum \Delta x = 1$, we assume $0 < \omega, l \leq 1$. Divide $[0, 1]$ into subintervals $[0, l/2], [l/2, 1]$; choose $N \geq 3$ so that $1/N < l/2$; and define $\delta = 1/N^2$:

$$\begin{array}{c|ccc} & & & \\ \hline 0 & 1/N & 1/2 & 1 \end{array}$$

Because f has $N - 1$ "jumps" on $[1/N, 1]$, we have fewer than N subintervals on $[l/2, 1]$ where f jumps by 1. Since each subinterval has length less than $\delta = 1/N^2$, the sum of the lengths of the subintervals with $\sup f - \inf f > \omega$ is less than $N(1/N^2) = 1/N < l/2$.

Of course we have an infinite number of jumps on $[0, l/2]$, but the total length of such subintervals cannot exceed $l/2$. Thus, the total length of subintervals, for any partition of $[0, 1]$ with $\delta = 1/N^2$ having $\sup f - \inf f > \omega$, is less than l.

b. Use Theorem 3.3.1 to show f is Riemann integrable on $[0, 1]$. Hint: $1/N < \epsilon$,

$$P_\epsilon = \left\{ 0, \frac{1}{N} - \frac{1}{2N^2}, \frac{1}{N}, \frac{1}{N} + \frac{1}{2N^2}, \ldots, \frac{1}{3} + \frac{1}{2N^2}, \frac{1}{2} - \frac{1}{2N^2}, \frac{1}{2}, \frac{1}{2} + \frac{1}{2N^2}, 1 \right\}.$$

Exercise 3.4.2. Given

$$f(x) = \begin{cases} 0 & x = 0, 1, \text{ or irrational for } 0 \le x \le 1, \\ 1/q & x = p/q, \text{ for } p, q \text{ relatively prime numbers,} \end{cases}$$

with $f \ge \frac{1}{2}$ at one point, $f \ge \frac{1}{3}$ at three points, and $f \ge 1/n$ at most ? points.

a. Show that f is continuous at the irrational numbers and discontinuous at the rational numbers. Hint: $J = \{x \mid f(x) \ge \epsilon\}$ is a finite set. For α irrational, choose an interval about α that contains no points of J.

b. Using Riemann's Theorem, show that f is Riemann integrable on $[0, 1]$. Hint: Choose N so that $1/N < \omega$ and $1/N < l$. Define $\delta = 1/N^3$. The jumps of f exceed $1/N$ on fewer than $N^2/2$ intervals, each having length less than $1/N^3$. What is the value of R $\int_0^1 f(x)\,dx$? Despite a dense set of discontinuities, f is Riemann integrable.

Exercise 3.4.3. In this exercise we construct Riemann's example of a function with a dense set of discontinuities that is Riemann integrable. Let

$$f(x) = \begin{cases} x & -\frac{1}{2} < x < \frac{1}{2}, \\ 0 & x = \pm\frac{1}{2}, \end{cases}$$

be extended periodically by $f(x) = f(x + 1)$.

Riemann's function

$$R(x) \equiv \sum_{n=1}^{\infty} \frac{f(nx)}{n^2}.$$

R is bounded by $\sum_{n=1}^{\infty} 1/2n^2$.

Because $f(x)$ is discontinuous iff x is an odd multiple of $\frac{1}{2}$, $f(nx)$ is discontinuous iff nx is an odd multiple of $\frac{1}{2}$, that is, iff $x = (2i - 1)/2n$ is a rational number with an odd numerator and even denominator.

Restricting our attention to $[0, 1]$, we have discontinuities at

$$\frac{1}{2}; \frac{1}{4}, \frac{3}{4}; \frac{1}{6}, \frac{3}{6}, \frac{5}{6}; \ldots; \frac{1}{2n}, \frac{3}{2n}, \ldots, \frac{2n-1}{2n};$$

and so on. We have odd multiples of $\frac{1}{2}$, odd multiples of $\frac{1}{4}$, odd multiples of $\frac{1}{6}, \ldots$, odd multiples of $1/2n, \ldots$, a dense subset of $[0, 1]$. The set $\{x \mid x = (2i - 1)/2n; i, n = \pm 1, \pm 2, \ldots\}$ is a dense subset of the reals. Riemann's function has a dense set of discontinuities. Again, consider $[0, 1]$. (The periodic nature of f extends the discussion to the real line.)

a. Suppose x is an odd multiple of $\frac{1}{2}$. Then (kx) is an odd multiple of $\frac{1}{2}$ iff $k = 1, 3, 5, \ldots$. So $f(kx^+) - f(kx^-) = -\frac{1}{2} - \frac{1}{2} = -1$ iff $k = 1, 3, 5, \ldots$. Because we have uniform convergence (Weierstrass M-test),

$$R(x^+) - R(x^-) = -\frac{1}{1^2} - \frac{1}{3^2} - \frac{1}{5^2} - \cdots = -\frac{\pi^2}{8}.$$

b. Suppose x is an odd multiple of $\frac{1}{4}$. Then (kx) is an odd multiple of $\frac{1}{2}$ iff $k = 2 \cdot 1, 2 \cdot 3, 2 \cdot 5, \ldots$. So $f(kx^+) - f(kx^-) = -1$ iff $k = 2, 6, 10, \ldots$.

$$R(x^+) - R(x^-) = -\frac{1}{2^2} - \frac{1}{6^2} - \frac{1}{10^2} - \cdots = -\frac{1}{2^2} \cdot \frac{\pi^2}{8}.$$

c. Suppose x is an odd multiple of $\frac{1}{6}$. Then (kx) is an odd multiple of $\frac{1}{2}$ iff $k = 3 \cdot 1, 3 \cdot 3, 3 \cdot 5, \ldots$. So $f(kx^+) - f(kx^-) = -1$ iff $k = 3, 9, 15, \ldots$.

$$R(x^+) - R(x^-) = -\frac{1}{3^2} - \frac{1}{9^2} - \frac{1}{15^2} - \cdots = -\frac{1}{3^2} \cdot \frac{\pi^2}{8}.$$

d. Suppose x is an odd multiple of $1/2n$. Then (kx) is an odd multiple of $\frac{1}{2}$ iff $k = n \cdot 1, n \cdot 3, n \cdot 5, \ldots$. So $f(kx^+) - f(kx^-) = -1$ iff $k = n, 3n, 5n, \ldots$.

$$R(x^+) - R(x^-) = -\frac{1}{n^2} - \frac{1}{3^2 n^2} - \frac{1}{5^2 n^2} - \cdots = -\frac{1}{n^2} \left(\frac{\pi^2}{8} \right).$$

In summary, R is a bounded function with a dense set of discontinuities.

e. Show R is Riemann integrable on $[0, 1]$. Hint: The interval $[0, 1]$ has a finite number of points (rational numbers of the form $(2i - 1)/2n$) where R "jumps" by $\pi/8n^2 > \omega$: $0 \le r_1 < r_2 < \cdots < r_K \le 1$.

We have shown that continuous functions are Riemann integrable. These last exercises show that, in some cases, a countable number of discontinuities may not prevent Riemann integrability.

3.5 Monotonic Functions Are Riemann Integrable

We begin by reviewing some characteristics of monotonic functions pertinent to Riemann integrability.

Exercise 3.5.1. Show that monotonic functions have right- and left-hand limits at each point of their domain.

Exercise 3.5.2. Show that monotonic functions have a countable number of discontinuities. Hint: $f(x^-) < r < f(x^+)$, for r a rational number. Noting that $\sum_P (\sup f - \inf f) \Delta x \le |f(b) - f(a)| \delta$, conclude the Riemann integrability of monotone functions.

In spite of this relatively straightforward argument for integrability, monotonic functions may be very interesting.

For the following exercises, let $r_1, r_2, r_3, \ldots$ be any enumeration of the rational numbers in $(0, 1)$. Define f on $[0, 1]$ by $f(x) = \sum_{\{n | r_n < x\}} 1/2^n$, where $0 < x < 1$, $f(0) = 0$, and $f(1) = 1$. Clearly f is monotone increasing and thus Riemann integrable on $[0, 1]$.

Exercise 3.5.3. Show that $\lim_{x \to r_K^-} f(x) = f(r_K^-) = f(r_K)$. Hint: Fix a rational number r_K in $(0, 1)$. Since monotonic functions have right- and left-hand limits, certainly $f(r_K^-) \le f(r_K)$. If $f(r_K^-) < f(r_K)$, choose $\epsilon > 0$ so that $f(r_K^-) + \epsilon < f(r_K)$, and then choose a natural number N so that $1/2^N < \epsilon/2$. Now select $\delta > 0$ so that $(r_K - \delta, r_K) \cap \{r_1, r_2, \ldots, r_N\} = \phi$. For $r_K - \delta < x < r_K$,

$$f(r_K) - f(x) = \sum_{\{n | x \le r_n < r_K\}} \frac{1}{2^n} < \frac{1}{2^{N+1}} + \cdots = \frac{1}{2^N} < \frac{\epsilon}{2}, \quad \text{or}$$

$$f(r_K) - \frac{\epsilon}{2} < f(x) \le f(r_K^-) < f(r_K) - \epsilon,$$

and we have a contradiction.

Exercise 3.5.4. Show that $f(r_k) + 1/2^k = f(r_k^+)$ for all k. Hint: For $r_k < x$, $f(x) - f(r_k) = \sum_{\{n | r_k \le r_n < x\}} 1/2^n$ and

$$\lim_{x \to r_k^+} f(x) = f(r_k) + \lim_{x \to r_k^+} \sum_{\{n | r_k \le r_n < x\}} \frac{1}{2^n} = f(r_k) + \frac{1}{2^k}.$$

That is, f has a jump of $1/2^k$ at each rational r_k.

Exercise 3.5.5. Show that f is continuous on the irrationals. Hint: Suppose α is an irrational number in $(0, 1)$, and let $\epsilon > 0$ be given. Choose N so that $1/2^N < \epsilon$. Construct an open interval about α that does not contain $r_1, r_2, \ldots, r_N$. Every rational in this subinterval has a subscript larger than N. If x is a point of this interval, $|f(x) - f(\alpha)| < \sum_{N+1} 1/2^n < \epsilon$, and f is a strictly increasing function that is continuous on the irrationals and has a jump of $1/2^k$ at r_k.

Again, a countable number of discontinuities does not prevent Riemann integrability.

3.6 Lebesgue's Criteria

Finally, in 1902, the French mathematician Henri Lesbesgue determined necessary and sufficient conditions for a bounded function to be Riemann integrable. (See Section 6.1.2.)

Theorem 3.6.1 (Lebesgue, 1902). *Suppose f is a bounded function of $[a, b]$. Then f is Riemann integrable on $[a, b]$ iff f is continuous almost everywhere.*

Proof. To say f is continuous almost everywhere on $[a, b]$ means we can "cover" the discontinuities of f with a countable collection of open intervals $I_1, I_2, \ldots$, whose length $\sum \ell(I_k)$ can be made arbitrarily small — that is, a set of measure zero.

The idea here is that where f is continuous ("most" of the interval $[a, b]$ by length), $\sup f - \inf f$ will be small. Where f is discontinuous such intervals comprise a "small" subset of $[a, b]$. Integrability is determined by the behavior of f on "most" of the interval.

We begin with the assumption that the bounded function f is Riemann integrable on $[a, b]$. We will show that the set of discontinuities of f is a set of measure zero. Show the set of discontinuities of f is the union of the sets

$$\left\{ x \in [a, b] \mid \lim_{\delta \to 0} \left(\sup_{(x-\delta, x+\delta)} f(z) - \inf_{(x-\delta, x+\delta)} f(z) \right) \geq \frac{1}{n} \right\},$$

for $n = 1, 2, \ldots$. It will be sufficient to show that such sets have measure zero.

Let $\epsilon > 0$ be given and fix n. From Cauchy's Criteria for Riemann integrability (Theorem 3.3.1) we have a partition P of $[a, b]$ so that

$$\sum_P (\sup f - \inf f) \, \Delta x < \frac{\epsilon}{4n}.$$

If the subinterval $[x_{k-1}, x_k]$ contains a point d with

$$\lim_{\delta \to 0} \left(\sup_{(d-\delta, d+\delta)} f(x) - \inf_{(d-\delta, d+\delta)} f(x) \right) \geq \frac{1}{n}$$

in its interior, then

$$\sup_{[x_{k-1},x_k]} f - \inf_{[x_{k-1},x_k]} f \geq \frac{1}{2n} \qquad \text{and} \qquad (\sup f - \inf f)\,\Delta x \geq \frac{\Delta x}{2n}$$

on this subinterval. But

$$\sum_{\text{``interior''}} (\sup f - \inf f)\,\Delta x + \sum_{\text{otherwise}} (\sup f - \inf f)\,\Delta x < \frac{\epsilon}{4n}.$$

We may conclude that $\sum_{\text{``interior''}} \Delta x < \epsilon/2$. Otherwise, such points are division points — a finite set — and can be covered by a finite number of intervals whose total length is less than $\epsilon/2$.

We have covered the set

$$\left\{ d \in [a,b] \mid \lim_{\delta \to 0} \left(\sup_{(d-\delta,d+\delta)} f - \inf_{(d-\delta,d+\delta)} f \right) \geq \frac{1}{n} \right\}$$

with a finite set of intervals of total length less than ϵ. By the arbitrary nature of ϵ, we conclude that this set and the points of discontinuity of the bounded function f have measure zero.

For the other direction, we assume the set of discontinuities of the bounded function f has measure zero (f is continuous almost everywhere). We will show that f is Riemann integrable on $[a,b]$.

We have a cover of the discontinuities of f by open intervals $I_1, I_2, \ldots$, with $\sum \ell(I_k) < \epsilon$. At each point x of continuity of f, we have an open interval J_x containing x so that $\sup f - \inf f$ over the closure of this open interval is less than ϵ. The collection of open intervals $I_1, I_2, \ldots, J_x$ is an open cover of the compact set $[a,b]$.

By the Heine–Borel Theorem, we have a finite subcollection that covers $[a,b]$. The endpoints of these subintervals (that are in $[a,b]$) and $\{a,b\}$ are division points for a partition P of $[a,b]$. Thus,

$$\sum_P (\sup f - \inf f)\,\Delta x < \left(\sup_{[a,b]} f - \inf_{[a,b]} f \right) \sum \ell(I_k) + \epsilon \sum \ell(J_x)$$

$$< \left(\sup_{[a,b]} f - \inf_{[a,b]} f \right) \epsilon + \epsilon(b - a).$$

Appealing to Cauchy's Criteria for Riemann Integrability (Theorem 3.3.1), we may conclude that f is Riemann integrable on $[a,b]$. The proof is complete. $\square$

The reader should compare this result with the corresponding result for Cauchy integrals (Theorem 2.2.1).

Exercise 3.6.1.

   a. Settle the Riemann integrability of the functions of Exercises 3.4.1, 3.4.2, and 3.4.3 using this result of Lebesgue.

   b. If f is Riemann integrable on $[a, b]$, then $|f|$ is Riemann integrable on $[a, b]$. Show that the converse is false. Hint: Let

$$f(x) = \begin{cases} 1 & x \text{ rational,} \\ -1 & x \text{ irrational,} \end{cases}$$

and consider the interval $[0, 1]$.

3.7   Evaluating à la Riemann

Now that we have existence of the Riemann integral for bounded, continuous almost everywhere functions, may we recover a function from its derivative using the Riemann integration process, as with the Cauchy integral? A Fundamental Theorem of Calculus offers some answers.

Theorem 3.7.1 (FTC for the Riemann Integral). *If F is a differentiable function on the interval $[a, b]$, and F' is bounded and continuous almost everywhere on $[a, b]$, then*

   1. *F' is Riemann integrable on $[a, b]$, and*

   2. *$\text{R} \int_a^x F'(t) \, dt = F(x) - F(a)$ for each x in the interval $[a, b]$.*

Proof. The first conclusion follows from Lebesgue's result (Theorem 3.6.1).

For the second part, existence of the Riemann integral of F' means that, given an $\epsilon > 0$, we have a partition P of $[a, x]$ so that $\sum_P \sup F' \, \Delta x$ and $\sum_P \inf F' \, \Delta x$ are within ϵ of each other. By the mean value theorem for derivatives,

$$F(x) - F(a) = \sum F(x_k) - F(x_{k-1}) = \sum_P F'(c_k) \, \Delta x.$$

Since $\sum_P F'(c_k) \, \Delta x$ is between $\sum_P \inf F' \, \Delta x$ and $\sum_P \sup F' \, \Delta x$, we have that $F(x) - F(a)$ is between $\sum_P \inf F' \, \Delta x$ and $\sum_P \sup F' \, \Delta x$, along with $\text{R} \int_a^x F'(t) \, dt$.

Thus $|F(x) - F(a) - \text{R} \int_a^x F'(t) \, dt| < 2\epsilon$. $\square$

Compare this with Theorem 2.3.1. The conditions on the derivative have been weakened from continuous to bounded and continuous almost everywhere. Look also at Theorems 6.4.2 and 8.7.3.

Exercise 3.7.1.

a. Given

$$F(x) = \begin{cases} x^2 \sin(\pi/x) & 0 < x \le 1, \\ 0 & x = 0, \end{cases}$$

show that R $\int_0^1 F'(x)\, dx = 0$.

b. Given

$$F(x) = \begin{cases} x^2 \sin(\pi/x^2) & 0 < x \le 1, \\ 0 & x = 0, \end{cases}$$

show that F' exists but F' is not Riemann integrable.

What are some of the properties of the Riemann integral? Can we recover a function from its integral using differentiation? Let's look at another Fundamental Theorem of Calculus for the Riemann integral.

Theorem 3.7.2 (Another FTC for the Riemann Integral). *Suppose f is a bounded and continuous almost everywhere function on the interval $[a, b]$. Define F on $[a, b]$ by $F(x) = R \int_a^x f(t)\, dt$.*

1. Then F is continuous (in fact, absolutely continuous) on $[a, b]$.

2. If f is continuous at a point x_0 in $[a, b]$, then F is differentiable at x_0 and $F'(x_0) = f(x_0)$.

3. $F' = f$ almost everywhere.

Proof. The function F is well defined by Lebesgue's result (Theorem 3.6.1). To show continuity of F, select B so that $|f| \le B$ on the interval $[a, b]$. Then

$$|F(y) - F(x)| = \left| R \int_x^y f(t)\, dt \right| \le B|y - x|.$$

Absolute continuity follows easily (see Definition 5.8.2).

Now we assume that f is continuous at x_0. Then given an $\epsilon > 0$, we have a $\delta > 0$ so that $f(x_0) - \epsilon < f(t) < f(x_0) + \epsilon$, for $t \in (x_0 - \delta, x_0 + \delta) \cap [a, b]$. Constants are Riemann integrable, and f is Riemann integrable by assumption.

From integration, then,

$$R \int_{x_0}^{x} [f(x_0) - \epsilon] \, dt \leq R \int_{x_0}^{x} f(t) \, dt$$

$$\leq R \int_{x_0}^{x} [f(x_0) + \epsilon] \, dt,$$

for $x \in [x_0, x_0 + \delta) \cap [a, b]$. Thus

$$-\epsilon \leq \frac{F(x) - F(x_0)}{x - x_0} - f(x_0) \leq \epsilon.$$

Argue $x \in (x_0 - \delta, x_0] \cap [a, b]$, and the conclusion follows.   □

Compare with Theorems 2.4.1, 6.4.1, and 8.8.1.

Exercise 3.7.2. Given $\ln x \equiv \int_1^x (1/t) \, dt$ for $x > 0$, show the following.

a. $(\ln x)' = 1/x$, for $x > 0$.

b. $\ln x < 0$ for $0 < x < 1$, $\ln 1 = 0$, and $\ln x > 0$ for $x > 1$.

c. $\ln(1/x) = -\ln x$.

d. $\ln(xy) = \ln x + \ln y$ for $x, y > 0$.

e. As $x \to \infty$, $\ln x \to \infty$. As $x \to 0^+$, $\ln x \to -\infty$.

f. As $x \to \infty$, $(\ln x)/x \to 0$. As $x \to 0^+$, $x \ln x \to 0$.

Exercise 3.7.3. Given

$$f(x) = \begin{cases} -1 & -1 \leq x \leq 0, \\ 0 & 0 < x < 1, \\ 1 & 1 \leq x \leq 3. \end{cases}$$

a. Calculate $R \int_{-1}^{x} f(t) \, dt$, $R \int_0^x f(t) \, dt$, $R \int_x^2 f(t) \, dt$, and $R \int_0^{2x} f(t) \, dt$.

b. Using the definition of continuity, investigate the continuity of these functions at $x = 0$ and $x = 1$. Calculate their derivatives where appropriate.

Exercise 3.7.4. Proceed as in the previous exercise, given

$$f(x) = \begin{cases} x & -1 \leq x \leq 1, \\ 1 & 1 < x < 2, \\ 3 - x & 2 \leq x \leq 3. \end{cases}$$

Exercise 3.7.5. Given

$$f(x) = \begin{cases} 0 & x = 0, \\ \sin(\pi/x) & 0 < x \le 1. \end{cases}$$

a. Sketch $F(x) = R \int_0^x f(t)\,dt$. Hint: $\int_0^x = \int_0^1 + \int_1^x$.

b. Is F continuous from the right at $x = 0$? Hint: $|F(x)| \le x$.

c. Is F differentiable from the right at $x = 0$? Hint: Write

$$R \int_0^x \sin\left(\frac{\pi}{t}\right) dt = R \int_0^x (-t^2) \sin\left(\frac{\pi}{t}\right) \left(-\frac{1}{t^2}\right) dt$$

and integrate by parts.

d. Suppose

$$f(x) = \begin{cases} 0 & x = 0, \\ \sin(\pi/x^\alpha) & 0 < x \le 1, \alpha > 0. \end{cases}$$

What now?

Exercise 3.7.6. Let $f(x) = -x$ for $-1 \le x < 1$ and $f(x+2) = 2f(x)$ for all x.

a. Calculate

$$F(x) = R \int_{-1}^x f(t)\,dt \qquad \text{for } t \text{ in } [-1, 4],$$

$$G(x) = R \int_{-1}^{2x} f(t)\,dt, \quad \text{and}$$

$$H(x) = R \int_{-1}^{1/2x} f(t)\,dt.$$

b. Discuss continuity and differentiability of F, G, and H.

3.8 Sequences of Riemann Integrable Functions

What about convergence theorems for sequences of Riemann integrable functions? We would like $\lim f_n = f$ to imply $\lim \int f_n = \int \lim f_n$. Let's explore this with some exercises.

Exercise 3.8.1. Given

$$f_k(x) = \begin{cases} kx & 0 \le x \le 1/k, \\ -k(x - 2/k) & 1/k \le x \le 2/k, \\ 0 & 2/k \le x \le 1. \end{cases}$$

Show that $\lim R \int_0^1 f_k(x)\, dx \ne R \int_0^1 (\lim f_k)(x)\, dx$.

Exercise 3.8.2. If $r_1, r_2, \ldots$ is an enumeration of the rational numbers in $(0, 1)$, define a sequence of functions $\{f_k\}$ by

$$f_k(x) = \begin{cases} 1 & x = r_1, r_2, \ldots, r_k, \\ 0 & \text{otherwise}, \end{cases} \qquad \text{for } 0 \le x \le 1.$$

Show that $\lim R \int_0^1 f_k(x)\, dx \ne R \int_0^1 (\lim f_k)(x)\, dx$, even though the sequence $\{f_k\}$ is monotonic and uniformly bounded.

In general,

$$\left| R \int_a^b f_k(x)\, dx - R \int_a^b f(x)\, dx \right| = \left| R \int_a^b [f_k(x) - f(x)]\, dx \right|$$

$$\le R \int_a^b |f_k(x) - f(x)|\, dx$$

$$\le (b - a) \sup_{[a,b]} |f_k - f| < \epsilon(b - a),$$

provided $|f_k - f| < \epsilon$ throughout the interval $[a, b]$. This would suggest uniform convergence — assuming, of course, the limit function f is Riemann integrable.

Theorem 3.8.1 (Convergence for Riemann Integrable Functions).   *If $\{f_k\}$ is a sequence of Riemann integrable functions converging uniformly to the function f on $[a, b]$, then f is Riemann integrable and $R \int_a^b f(x)\, dx = \lim R \int_a^b f_k(x)\, dx$.*

Proof. We will show that f is Riemann integrable on $[a, b]$. Let $\epsilon > 0$. From the uniform convergence of f_k on $[a, b]$, given $\epsilon > 0$, we have $K > 0$, so that

$$f_k(x) - \frac{\epsilon}{4(b - a)} \le f(x) \le f_k(x) + \frac{\epsilon}{4(b - a)} \qquad (1)$$

for all x in $[a, b]$, when $k \ge K$.

So f is bounded on $[a, b]$, and because f_k is Riemann integrable on $[a, b]$ we have a partition P of $[a, b]$ so that $\sum_P (\sup f_k - \inf f_k) \Delta x < \epsilon/2$ (see Theorem 3.3.1). However, from (3.1) we have

$$\sum_P \inf f_k \, \Delta x - \frac{\epsilon}{4} \le \sum_P \inf f \, \Delta x \le \sum_P \sup f \, \Delta x$$

$$\le \sum_P \sup f_k \, \Delta x + \frac{\epsilon}{4}, \qquad \text{for } k \ge K.$$

That is, $\sum_P (\sup f - \inf f) \Delta x < \epsilon$. We may conclude that f is Riemann integrable on $[a, b]$, and then by integrating (3.1), the second conclusion follows.   □

Exercise 3.8.3.

a. Given $f_k(x) = \dfrac{2x}{1 + k^3 x^3}$, for $0 \le x \le 1$, calculate

$$\lim_k R \int_0^1 \frac{2x}{1 + k^3 x^3} \, dx.$$

b. Calculate $R \int_0^1 (\text{Riemann's Function}) \, dx$.

c. Calculate the Riemann integral of the function in Exercise 3.4.1 by defining an appropriate sequence of functions $\{f_k\}$.

We are now in a position to give (in slightly modified form) an example due to Volterra (1881) of a function with a bounded derivative whose derivative is not Riemann integrable. This example prompted Lebesgue to develop an integration process so that such functions would be (Lebesgue) integrable. Gordon (1994) supplies a complete explanation in his book.

Our approach to the example has three stages: construction of the Cantor set, construction of a modified Cantor set, and a return to $x^2 \sin(1/x)$.

3.9   The Cantor Set (1883)

On a list of famous mathematical objects, the Cantor set would appear early. (I dare say we can all remember our first encounter with it.) It is the origin of many examples and counterexamples in analysis. We begin with these steps.

1. Divide $[0, 1]$ into three equal parts and remove the middle third, the open interval $(\frac{1}{3}, \frac{2}{3})$. We are left with the closed set $F_1 = [0, \frac{1}{3}] \cup [\frac{2}{3}, 1]$.

2. Divide each of the two closed subintervals of F_1 into three equal parts and remove the middle thirds, the open intervals $(\frac{1}{9}, \frac{2}{9})$ and $(\frac{7}{9}, \frac{8}{9})$. The closed set $F_2 = [0, \frac{1}{9}] \cup [\frac{2}{9}, \frac{1}{3}] \cup [\frac{2}{3}, \frac{7}{9}] \cup [\frac{8}{9}, 1]$ remains.

3. Continuing in this way, after n steps we have deleted $1 + 2^1 + \cdots + 2^{n-1} = 2^n - 1$ open intervals and have left the closed set F_n, consisting of 2^n closed intervals, each of length $1/3^n$.

The Cantor set C is the intersection of the closed sets F_n:

$$C = \cap F_n.$$

For comparison, here is an equivalent formulation: The Cantor set is the points of $[0, 1]$ that have a base 3 expansion without the digit 1. That is, for $x \in C$,

$$x = \frac{a_1}{3^1} + \frac{a_2}{3^2} + \frac{a_3}{3^3} + \cdots, \qquad \text{where } a_n = 0 \text{ or } 2.$$

Example 3.9.1. The Cantor set C has measure zero (Exercise 5.5.2). We will show that C is uncountable.

This statement may seem unreasonable. After all, the only obvious members of the Cantor set are $0, 1, \frac{1}{3}, \frac{2}{3}, \frac{1}{9}, \frac{2}{9}, \frac{7}{9}, \frac{8}{9}, \ldots$, the countable set of endpoints of the open intervals that we removed. But we have something besides endpoints in this set, for example

$$.020202 \cdots = \frac{2}{3^2} + \frac{2}{3^4} + \frac{2}{3^6} + \cdots = \frac{2/9}{1 - 1/9} = \frac{1}{4}.$$

It is appropriate that Cantor's diagonalization process can be used effectively. Assume that C is countable and make a list of base 3 expansions:

$$x_1 = .a_{11}a_{12}a_{13} \cdots$$
$$x_2 = .a_{21}a_{22}a_{23} \cdots$$
$$\vdots$$
$$x_n = .a_{n1}a_{n2}a_{n3} \cdots,$$
$$\vdots$$

with $a_{nm} = 0$ or 2. Let $x = .a_1a_2a_3 \ldots$, where

$$a_n = \begin{cases} 0 & a_{nn} = 2, \\ 2 & a_{nn} = 0. \end{cases}$$

The number x is not on our list. We have an uncountable set of measure zero.

The Cantor set, a closed set, contains no intervals. It is nowhere dense:

$$\left(\frac{3k+1}{3^n}, \frac{3k+2}{3^n}\right) \subset (a,b) \subset (0,1) \qquad \text{for } n \text{ sufficiently large.}$$

Every point of the Cantor set is a limit of points in C (so it is a perfect set) and points not in C. For example, suppose $x_0 \in C = \cap F_n$ and I_n is a closed interval of F_n that contains x_0. Since the length of I_n is $1/3^n$, we may choose n so large, say N, such that for any open interval (a,b) continuing x_0, $I_N \subset (a,b)$. At least one of the endpoints is different from x_0.

Thus, given any open interval about a point of the Cantor set, it contains a different point of the Cantor set and a point not in the Cantor set. The Cantor set is a *nowhere dense perfect set of measure zero*.

For Volterra's example, we want a nowhere dense perfect set of *positive* measure. We will construct a modified Cantor set where we do not remove as much at each stage.

3.10   A Nowhere Dense Set of Positive Measure

Our construction of a modified Cantor set occurs in a series of steps, as follows.

1. Let a_1 be positive and less than $\frac{1}{2}$. Remove the open interval $(a_1, 1 - a_1)$ from $[0,1]$. We are left with the closed sets $F_{1,1} : [0, a_1]$ and $F_{1,2} : [1 - a_1, 1]$, each of length a_1. See Figure 1(a).

2. Let a_2 be positive and less than $\frac{1}{2}a_1$. Remove the open intervals $(a_2, a_1 - a_2)$ and $(1 - a_1 + a_2, 1 - a_2)$. We are left with the closed sets $F_{2,1} = [0, a_2]$, $F_{2,2} = [a_1 - a_2, a_1]$, $F_{2,3} = [1 - a_1, 1 - a_1 + a_2]$, and $F_{2,4} = [1 - a_2, 1]$, each of length a_2. See Figure 1(b).

3. At step $n-1$ we have closed intervals $F_{n-1,1}, \ldots, F_{n-1,2^{n-1}}$, each of length a_{n-1}.

4. At step n each of these closed intervals $F_{n-1,k}$ is divided into two closed intervals and one open interval; see Figure 1(c). We have

$$I_{n,k} = F_{n-1,k} - (F_{n,2k-1} \cup F_{n,2k}),$$

and $I_{n,k}$ has length $a_{n-1} - 2a_n$.

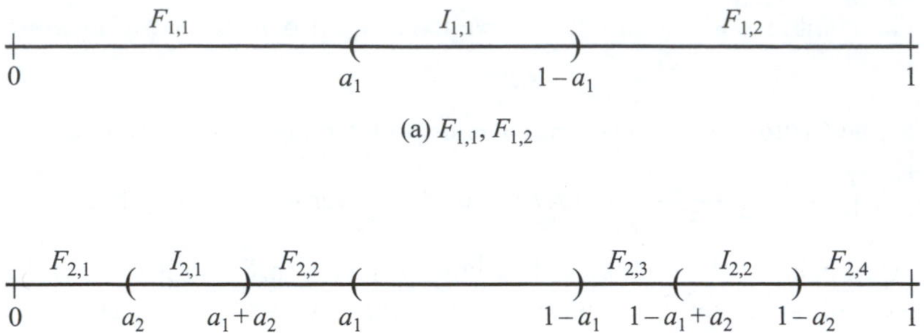

(a) $F_{1,1}, F_{1,2}$

(b) $F_{2,1}, F_{2,2}, F_{2,3}, F_{2,4}$

Figure 1. Constructing the modified Cantor set

What have we constructed? Let

$$C = [0,1] \cap (F_{1,1} \cup F_{1,2}) \cap (F_{2,1} \cup F_{2,2} \cup F_{2,3} \cup F_{2,4}) \cap \cdots.$$

Let $a_n = [(1/2^n) + (1/3^n)]/2$. Then

$$a_0 - 2a_1 = \frac{1}{6} \quad \text{and} \quad \ell(I_{1,1}) = \frac{1}{6},$$

$$a_1 - 2a_2 = \frac{1}{18} \quad \text{and} \quad \ell(I_{2,1}) + \ell(I_{2,2}) = 2\left(\frac{1}{18}\right),$$

$$a_2 - 2a_3 = \frac{1}{54} \quad \text{and} \quad \ell(I_{3,1}) + \ell(I_{3,2}) + \ell(I_{3,3}) + \ell(I_{3,4}) = 4\left(\frac{1}{54}\right), \ldots.$$

The sum of the lengths of the open intervals removed is $\frac{1}{6} + 2(\frac{1}{18}) + 4(\frac{1}{54}) + \cdots = \frac{1}{2}$. This modified Cantor set has measure $\frac{1}{2}$. In general,

$$1 - [(a_0 - 2a_1) + 2(a_1 - 2a_2) + 4(a_2 - 2a_3) + \cdots + 2^{n-1}(a_{n-1} - 2a_n)] = 1 - 2^n a_n.$$

We claim that this modified Cantor set is a nowhere dense perfect set of measure $\frac{1}{2}$. For suppose that $p \in C$. Since $\ell(F_{n,k}) = a_n \to 0$, given any neighborhood of p we have N and K so that $p \in F_{N,K} \subset (p - \delta, p + \delta)$. In fact, every neighborhood of p contains points of C and points not in C, namely those in $(I_{N+1,K})$. See Figure 2.

In the next section, we introduce the remarkable *Cantor function*, developed as the uniform limit of a sequence of continuous functions.

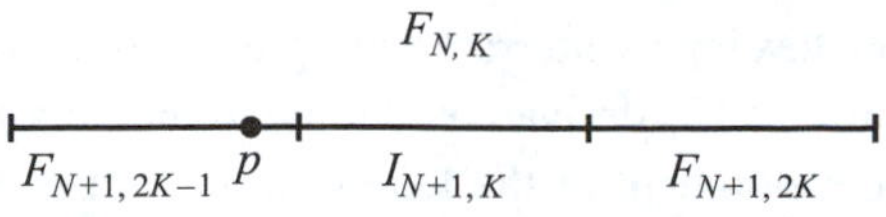

Figure 2. $F_{N,K}$

3.11   Cantor Functions

We will construct a sequence $\{c_n\}$ of continuous, monotone increasing functions on $[0, 1]$. The Cantor function will be the uniform limit of this sequence. Here are the steps:

1. $c_1(0) = 0$, $c_1(x) = \frac{1}{2}$, $\frac{1}{3} \leq x \leq \frac{2}{3}$, $c_1(1) = 1$, $c_1(x)$ is linear otherwise and continuous.

2. $c_2(0) = 0$, $c_2(x) = \frac{1}{4}$ if $\frac{1}{9} \leq x \leq \frac{2}{9}$, $\frac{2}{4}$ if $\frac{1}{3} \leq x \leq \frac{2}{3}$, $\frac{3}{4}$ if $\frac{7}{9} \leq x \leq \frac{8}{9}$, $c_2(1) = 1$, $c_2(x)$ linear otherwise and continuous.

3. After n steps we have the deleted $2^n - 1$ open intervals. On them c_n assumes the values $1/2^n, 2/2^n, \ldots, (2^n - 1)/2^n$ left to right. $c_n(0) = 0$, $c_n(1) = 1$, and $c_n(x)$ is linear on the 2^n closed intervals and continuous.

The Cantor function C is the pointwise limit of this sequence of increasing continuous functions. Clearly $C(0) = 0$, $C(1) = 1$. We claim C is continuous. Show that

$$|c_1 - c_2| \leq \frac{1}{4} - \frac{3}{4}\left(\frac{1}{4}\right),$$

$$|c_2 - c_3| \leq \frac{1}{8} - \left(\frac{3}{2}\right)^2 \frac{1}{27},$$

$$\cdots$$

$$|c_n - c_{n+1}| \leq \frac{1}{2^{n+1}} - \left(\frac{3}{2}\right)^n \left(\frac{1}{3^{n+1}}\right) = \frac{1}{3 \cdot 2^{n+1}}.$$

Consequently, $\lim c_n = C$ uniformly on $[0, 1]$. Furthermore, $C' = 0$ on the intervals $\left(\frac{1}{3}, \frac{2}{3}\right)$, $\left(\frac{1}{9}, \frac{2}{9}\right) \cup \left(\frac{7}{9}, \frac{8}{9}\right) \cup \cdots$, a set of measure 1.

The continuous function C is 0 at 0, 1 at 1, but has a derivative equal to 0 almost everywhere. C is a monotone function on $[0, 1]$ whose derivative vanishes almost everywhere.

Exercise 3.11.1. Calculate the integrals $C \int_0^1 C(x)\, dx$ and $C \int_0^1 xC(x)\, dx$. Hint: Theorem 3.8.1.

The next section develops Volterra's example of a function with a bounded but not Riemann integrable derivative. As mentioned before, Volterra's example prompted development of the Lebesgue integral.

3.12   Volterra's Example

Vito Volterra's example dates from 1881.

Let C denote a Cantor set of positive measure, say $\frac{1}{2}$. Suppose $I_{n,k} = (a, b)$ is one of the open intervals removed in the construction process for C. Let

$$c = \sup \left\{ x \ \middle| \ a < x \leq \frac{a+b}{2}, \left[(x-a)^2 \sin \left(\frac{1}{x-a} \right) \right]' = 0 \right\},$$

and define a function $f_{n,k}$ on $I_{n,k} = (a, b)$ by

$$f_{n,k}(x) = \begin{cases} (x-a)^2 \sin \left(\frac{1}{x-a} \right) & a < x \leq c, \\ (c-a)^2 \sin \left(\frac{1}{c-a} \right) & c \leq x \leq b+a-c, \\ (x-b)^2 \sin \left(\frac{1}{b-x} \right) & b+a-c < b. \end{cases}$$

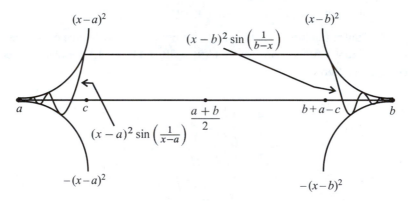

Figure 3. $f_{n,k}$

The function $f_{n,k}$ is differentiable on $I_{n,k}$, the derivative is bounded by 3 on $I_{n,k}$, and

$$f'_{n,k}\left(a + \frac{1}{n\pi}\right) = f'_{n,k}\left(b - \frac{1}{n\pi}\right) = \pm 1.$$

We define Volterra's function V on $[0, 1]$ as follows:

$$V(x) = \begin{cases} f_{n,k}(x) & x \in I_{n,k}; n, k = 1, 2, \ldots, \\ 0 & \text{otherwise}. \end{cases}$$

We claim function V is differentiable on $[0, 1]$. Certainly the open intervals $I_{n,k}$ do not pose a problem. Suppose x_0 belongs to C and $x > x_0$. If x also belongs to C, then $V(x) = V(x_0) = 0$. Otherwise, x belongs to some $I_{n,k} = (a, b)$, and $x_0 < a < x < b$. But

$$V(x) - V(x_0) = V(x) \leq \max\{(x - a)^2, (b - x)^2\} \leq (x - x_0)^2$$

and

$$\frac{V(x) - V(x_0)}{x - x_o} \leq x - x_0.$$

Thus $V' = 0$ on C. So V is differentiable on $[0, 1]$, and $|V'| \leq 3$ on $[0, 1]$. We claim V' is not continuous on C, a set of positive measure $\frac{1}{2}$, and thus is not Riemann integrable on the interval $[0, 1]$.

Let $x_0 \in C$. If x_0 is an endpoint of some $I_{n,k} = (a, b)$, $x_0 = a$ or b, we have a sequence of points $\{a + 1/n\pi\}$ or $\{b - 1/n\pi\}$ with $V' = \pm 1$, whereas $V'(a) = V'(b) = 0$.

If $x_0 \in C$ and x_0 is not an endpoint of $I_{n,k}$, then we have a sequence of endpoints converging to x_0 (nowhere dense), and thus x_0 will again have points arbitrarily close to x_0 with $V' = \pm 1$, whereas $V'(x_0) = 0$.

This concludes Volterra's example.

So the Riemann integral does not necessarily integrate *bounded* derivatives. We will show in Chapter 6 that the Lebesgue integral does not have this defect. In fact, $L \int_0^1 F'(x)\, dx = F(1) - F(0)$ in this example.

3.13 Lengths of Graphs and the Cantor Function

Does the graph of the Cantor function have a length? The Cantor function has a nice graph, continuous and nondecreasing. The length would be

$$\sup_P \left\{ \sum \left[(x_k - x_{k-1})^2 + [C(x_k) - C(x_{k-1})]^2 \right]^{1/2} \right\},$$

where P is any partition of $[0, 1]$. Because

$$\sum_P \left[(\Delta x)^2 + (\Delta C)^2 \right]^{1/2} \leq \sum (\Delta x + \Delta C) = 2$$

for any partition, the length of the graph of C is at most 2.

We claim that the length of the graph of C is actually 2. Let's analyze this claim.

1. At the first stage of the process, we have a horizontal line segment of length $\frac{1}{3}$, and what remains exceeds $2(\frac{1}{2})$: Use the partition $P_1 = \{0, \frac{1}{3}, \frac{2}{3}, 1\}$.

2. At the second stage, we have three horizontal line segments of length $\frac{1}{3} + 2(\frac{1}{9})$, and what remains exceeds $4(\frac{1}{4})$: $P_2 = \{0, \frac{1}{9}, \frac{2}{9}, \frac{1}{3}, \frac{2}{3}, \frac{7}{9}, \frac{8}{9}, 1\}$ is the appropriate partition.

3. At the third stage, we have seven horizontal line segments of length $\frac{1}{3} + 2(\frac{1}{9}) + 4(\frac{1}{27})$ and what remains exceeds $8(\frac{1}{8})$, and so on.

4. Thus, at the nth stage, the length of the graph of the Cantor function exceeds $(\frac{1}{3}) + 2(\frac{1}{9}) + 4(\frac{1}{27}) + \cdots + 2^{n-1}(1/3^n) + 2^n(1/2^n)$, and this quantity converges to 2.

A complete description of the Cantor function and other remarkable functions may be found in the beautiful books by Sagan (1994) and by Kannan and Krueger (1996).

3.14 Summary

Two Fundamental Theorems of Calculus for the Riemann Integral

If F is differentiable on $[a, b]$, and F' is bounded and continuous almost everywhere on $[a, b]$, then

1. F' is Riemann integrable on $[a, b]$, and

2. $\text{R} \int_a^x F'(t)\, dt = F(x) - F(a)$, where $a \leq x \leq b$.

If f is bounded and continuous almost everywhere on $[a, b]$ and $F(x) = \text{R} \int_a^x f(t)\, dt$, then

1. F is absolutely continuous on $[a, b]$,

2. F is differentiable almost everywhere on $[a, b]$, and

3. $F' = f$ at points of continuity of f (almost everywhere).

Figure 4. Cauchy integrability implies Riemann integrability

3.15   References

1. Bressoud, David. *A Radical Approach to Real Analysis.* Washington: Mathematical Association of America, 1994.

2. Darboux, J. Gaston. Mémoire sur les fonctions discontinues. *Annales Sci. École Normale Supérieure* 4:2 (1875) 57–112.

3. Goldberg, Richard R. *Methods of Real Analysis.* New York: Wiley, 1964.

4. Gordon, Russell. *The Integrals of Lebesgue, Denjoy, Perron and Henstock.* Providence, R.I.: American Mathematical Society, 1994.

5. Kannan, Rangachary, and Carole King Krueger. *Advanced Analysis on the Real Line.* New York: Springer, 1996.

6. Ross, Kenneth. *Elementary Analysis: The Theory of Calculus.* New York: Springer, 1980.

7. Sagan, Hans. *Space-Filling Curves.* New York: Springer, 1994.

8. Simmons, George. *Calculus Gems.* New York: McGraw-Hill, 1992.

9. Stromberg, Karl. *Introduction to Classical Real Analysis.* Belmont, Calif.: Wadsworth, 1981.

CHAPTER 4

The Riemann–Stieltjes Integral

Neglect of mathematics works injury to all knowledge, since he who is ignorant of it cannot know the other sciences or things of this world. And what is worse, men who are thus ignorant are unable to perceive their own ignorance and so do not seek a remedy. — Roger Bacon

4.1 Generalizing the Riemann Integral

After Riemann's formulation of the integral, various generalizations were attempted. One of the most successful, the so-called *Riemann–Stieltjes integral*, was obtained by Thomas Stieltjes (1856–1894). Stieltjes was trying to model mathematically the physical problem of computing moments for various mass distributions on the x-axis, with masses m_i at distances d_i from the origin (Birkhoff, 1973):

"Such a distribution will be perfectly determined if one knows how to calculate the total mass distributed over each segment Ox [of OX]. This evidently will be an increasing function of x.... Accordingly, let $\phi(x)$ be an increasing function defined on the interval (a, b).... It is convenient always to regard $\phi(b) - \phi(a)$ as the mass contained in the interval (a, b).... Let us now consider the moment of such a mass distribution wrt [with respect to] the origin. Let us set $a = x_0$, $b = x_n$, and let us intercalate $n - 1$ values between x_0 and x_n:

$$x_0 < x_1 < x_2 < \cdots < x_{n-1} < x_n.$$

Next, let us take n numbers $\xi_1, \xi_2, \ldots, \xi_n$ such that $x_{k-1} \leq \xi_k \leq x_k$. The limit of the sum $\xi_1 [\phi(x_1) - \phi(x_0)] + \xi_2 [\phi(x_2) - \phi(x_1)] + \cdots + \xi_n [\phi(x_n) - \phi(x_{n-1})]$ will be the moment, by definition. More

75

generally, let us consider the sum

$$f(\xi_1)\,[\phi(x_1) - \phi(x_0)] + f(\xi_2)\,[\phi(x_2) - \phi(x_1)]$$
$$+ \cdots + f(\xi_n)\,[\phi(x_n) - \phi(x_{n-1})]\,.$$

This will still have a limit, which we shall denote by $\int_a^b f(x)\,d\phi(x)$."

So calculation of moments for mass distributions has led us to consider sums, *Riemann–Stieltjes sums*, of the form

$$\sum_{k=1}^{n} f(c_k)\,[\phi(x_k) - \phi(x_{k-1})] = \sum_P f\Delta\phi.$$

These are Riemann sums when $\phi(x) = x$.

The optimism expressed by Stieltjes — "this will still have a limit" — is what we want to investigate. But first, a more formal definition is needed.

Definition 4.1.1 (The Riemann–Stieltjes Integral). (Stieltjes, 1894) Suppose that we have two bounded functions f and ϕ defined on the interval $[a, b]$, and we have a number A such that for each $\epsilon > 0$ there exists a positive constant δ for which

$$\left| \sum_{k=1}^{n} f(c_k)\,[\phi(x_k) - \phi(x_{k-1})] - A \right| < \epsilon, \qquad \text{where } x_{k-1} \le c_k \le x_k,$$

for every partition P of $[a, b]$ whose subintervals have length less than δ. We say f is Riemann–Stieltjes integrable with respect to ϕ on $[a, b]$, and we write

$$\text{R--S}\int_a^b f(x)\,d\phi(x) = A.$$

Exercise 4.1.1.

a. Suppose f is constant on $[a, b]$ with $f(x) = k$. Then for any partition of $[a, b]$,

$$\sum_P f\Delta\phi = k\,[\phi(b) - \phi(a)]. \quad \text{R--S}\int_a^b k\,d\phi(x) = k\,[\phi(b) - \phi(a)].$$

b. Suppose ϕ is constant on $[a, b]$ with $\phi(x) = k$. Then for any partition of $[a, b]$,

$$\sum_P f\Delta\phi = 0. \quad \text{R--S}\int_a^b f(x)\,d\phi(x) = 0.$$

4.2 Discontinuities

Thinking back to the mass distribution problem that Stieltjes was trying to solve, and the possibility of point masses, suppose that ϕ has a jump discontinuity, and f is "nice."

Exercise 4.2.1. Assume f is continuous on $[a, b]$ and

$$\phi(x) = \begin{cases} \beta_1 & a \leq x < \hat{x}, \\ \beta & x = \hat{x}, \\ \beta_2 & \hat{x} < x \leq b. \end{cases}$$

That is, ϕ is a step function with one step at $\hat{x}$, for $a < \hat{x} < b$. Let P be a partition of $[a, b]$.

  a. What happens if $\hat{x}$ is a partition point, say $\hat{x} = x_K$ for $0 < K < n$? In this case, $x_{K-1} \leq c_K \leq \hat{x} = x_K \leq c_{K+1} \leq x_{K+1}$. Show

$$\sum_P f \Delta\phi = [f(c_K) - f(\hat{x})] (\beta - \beta_1)$$

$$+ [f(c_{K+1}) - f(\hat{x})] (\beta_2 - \beta) + f(\hat{x})(\beta_2 - \beta_1).$$

  Hint: As the subintervals of P become small, the continuity of f makes the first two terms approach zero, suggesting that the Riemann–Stieltjes integral has the value $f(\hat{x})(\beta_2 - \beta_1) = f(\hat{x}) \left[\phi(\hat{x}^+) - \phi(\hat{x}^-) \right]$.

  b. What if $\hat{x}$ is not a partition point? Hint: We have $x_{K-1} < \hat{x} < x_K$ and $x_{K-1} \leq c_K \leq x_K$. Show

$$\sum_P f \Delta\phi = [f(c_K) - f(\hat{x})] (\beta_2 - \beta_1) + f(\hat{x})(\beta_2 - \beta_1);$$

  and the continuity of f suggests that the R-S integral has the value $f(\hat{x})[\phi(\hat{x}^+) - \phi(\hat{x}^-)]$.

  c. Show that R-S $\int_a^b f(x)\, d\phi(x) = f(\hat{x})[\phi(\hat{x}^+) - \phi(\hat{x}^-)]$. Generalize to $\hat{x} = a$, $\hat{x} = b$, and n steps for ϕ.

The preceding exercise results in the following theorem.

Theorem 4.2.1. *Suppose that f is continuous on $[a, b]$ and that ϕ is a step function with jumps at $\hat{x}_1, \hat{x}_2, \ldots, \hat{x}_n$ in $[a, b]$. Then the Riemann–Stieltjes integral of f with respect to ϕ exists, and*

$$\text{R-S} \int_a^b f(x)\, d\phi(x) = \sum_{k=1}^n f(\hat{x}_k)[\phi(\hat{x}_k^+) - \phi(\hat{x}_k^-)],$$

where $\phi(a^+) - \phi(a^-)$ means $\phi(a^+) - \phi(a)$, and $\phi(b^+) - \phi(b^-)$ means $\phi(b) - \phi(b^-)$.

Exercise 4.2.2. Suppose that $f(x) = x^2$ and $\phi(x) = [x]$, the *greatest integer function*, where $0 \le x \le n$. Show that

$$\text{R-S} \int_0^n x^2 d([x]) = 1^2 + 2^2 + \cdots + n^2 = \frac{n(n+1)(2n+1)}{6}.$$

What happens if f and ϕ have a common point of discontinuity? If

$$f(x) = \begin{cases} 1 & x \ne \frac{1}{2}, \\ 2 & x = \frac{1}{2} \end{cases} \quad \text{and} \quad \phi(x) = \begin{cases} 1 & 0 \le x < \frac{1}{2}, \\ 3 & \frac{1}{2} \le x \le 1, \end{cases}$$

both are discontinuous at $x = \frac{1}{2}$. Suppose P is a partition of $[0, 1]$ and $\frac{1}{2}$ is the partition point x_k. Then

$$\sum_P f\Delta\phi = f(c_k)\left[\phi\left(\frac{1}{2}\right) - \phi(x_{k-1})\right]$$

$$+ f(c_{k+1})\left[\phi(x_{k+1}) - \phi\left(\frac{1}{2}\right)\right] = 2f(c_k).$$

If $\frac{1}{2}$ is not a partition point, $\sum_P f\Delta\phi = 2f(c_k)$. Whether $\frac{1}{2}$ is a partition point or not, the result is the same.

We have problems since $f(c_k)$ could be 2 or 1 according as c_k is $\frac{1}{2}$ or is different from $\frac{1}{2}$. The Riemann–Stieltjes integral does not exist. So, in all that follows, we will be assuming that f and ϕ are real-valued bounded functions on $[a, b]$ with no common point of discontinuity.

Exercise 4.2.3. Suppose $f(x) = x$ and $\phi(x) = x^2$ on $[0, 1]$. Let P be any partition of $[0, 1]$, and form the Riemann–Stieltjes sum $\sum_P f(c)\Delta\phi = \sum_{k=1}^n c_k\left[x_k^2 - x_{k-1}^2\right]$. Then,

$$\sum x_{k-1}[x_k^2 - x_{k-1}^2] \le \sum c_k[x_k^2 - x_{k-1}^2] \le \sum x_k[x_k^2 - x_{k-1}^2].$$

Subtracting $\sum \frac{2}{3}[x_k^3 - x_{k-1}^3]$ from each term, we have

$$\left|\sum c_k[x_k^2 - x_{k-1}^2] - \frac{2}{3}\right| \le \sum (x_k - x_{k-1})^2 \le \delta.$$

We conclude R–S $\int_0^1 x d(x^2) = \frac{2}{3} = $ R $\int_0^1 x \, 2x \, dx$.

You may wonder: where did $\sum \frac{2}{3}[x_k^3 - x_{k-1}^3]$ come from? We have "fundamental theorems" relating Riemann–Stieltjes integrals to Riemann integrals.

4.3   Existence of Riemann–Stieltjes Integrals

Theorem 4.3.1. *Suppose f is continuous and ϕ is differentiable, with ϕ' being Riemann integrable, on the interval $[a, b]$. Then the Riemann–Stieltjes integral of f with respect to ϕ exists, and*

$$\text{R–S} \int_a^b f(x)\, d\phi(x) = \text{R} \int_a^b f(x)\phi'(x)\, dx.$$

Proof. Since $f\phi'$ is Riemann integrable, we are trying to show that the number A as defined earlier (Section 4.1.1) equals $\text{R} \int_a^b f(x)\phi'(x)\, dx$. An estimate for $\sum_P f\Delta\phi - \text{R} \int_a^b f(x)\phi'(x)\, dx$ is needed.

However, with the observation that $\Delta\phi = \text{R} \int_{x_{k-1}}^{x_k} \phi'(x)\, dx$, by the Fundamental Theorem of Calculus for the Riemann Integral (Theorem 3.7.1), we have that, for any partition P,

$$\sum_P f\Delta\phi - \text{R} \int_a^b f(x)\phi'(x)\, dx = \sum_{k=1}^n \text{R} \int_{x_{k-1}}^{x_k} [f(c_k) - f(x)]\phi'(x)\, dx.$$

The uniform continuity of f and the boundedness of ϕ' will now complete the argument.   □

Exercise 4.3.1.

  a. Looking at Exercise 4.2.3, calculate $\text{R–S} \int_0^1 x\, d(x^2)$ in light of the theorem we just proved.

  b. Calculate $\text{R–S} \int_{-1}^1 x\, d(x^2)$.

  c. Calculate $\text{R–S} \int_{-1}^1 |x|\, d(x\, |x|)$.

  We have a theorem that can be helpful in some evaluations.

Theorem 4.3.2. *Suppose f and ϕ are bounded functions with no common discontinuities on the interval $[a, b]$, and the Riemann–Stieltjes integral of f with respect to ϕ exists. Then the Riemann–Stieltjes integral of ϕ with respect to f exists, and*

$$\text{R–S} \int_a^b \phi(x)\, df(x) = f(b)\phi(b) - f(a)\phi(a) - \text{R–S} \int_a^b f(x)\, d\phi(x).$$

Proof. A partition of $[a, b]$ with cs between xs may be viewed as a partition of $[a, b]$ with xs between cs.

Let $\epsilon > 0$ be given. Since by assumption f is Riemann–Stieltjes integrable with respect to ϕ, we have a positive δ so that, for any partition of

$[a, b]$ whose subintervals have length less than δ, the associated Riemann–Stieltjes sum is within ϵ of R–S $\int_a^b f(x) \, d\phi(x)$.

Let P be a partition of $[a, b]$ whose subintervals have length less than $\delta/2$. If $c_0 = a$ and $c_{n+1} = b$, then $\{c_0, c_1, c_2, \ldots, c_n, c_{n+1}\}$ is a partition of $[a, b]$, for $c_k \leq x_k \leq c_{k+1}$, and $c_k - c_{k-1} < \delta$.

Then

$$
\left| \sum_{k=1}^n \phi(c_k)[f(x_k) - f(x_{k-1})] \right.
$$

$$
\left. - \left(f(b)\phi(b) - f(a)\phi(a) - \text{R–S} \int_a^b f(x) d\phi(x) \right) \right|
$$

$$
= \left| \sum_{k=1}^n \phi(c_k)[f(x_k) - f(x_{k-1})] \right.
$$

$$
\left. -[f(x_n)\phi(c_{n+1}) - f(x_0)\phi(c_0)] + \text{R–S} \int_a^b f(x) d\phi(x) \right|
$$

$$
= \left| \sum_{k=0}^n \phi(c_k) f(x_k) - \sum_{k=1}^{n+1} \phi(c_k) f(x_{k-1}) + \text{R–S} \int_a^b f(x) d\phi(x) \right|
$$

$$
= \left| \text{R–S} \int_a^b f(x) d\phi(x) - \sum_{k=0}^n f(x_k)[\phi(c_{k+1}) - \phi(c_k)] \right| < \epsilon,
$$

because the Riemann–Stieltjes sum $\sum f(x_k) \Delta\phi$ is associated with a partition of $[a, b]$ whose subintervals, $[c_k, c_{k+1}]$, have length less than δ. $\square$

Exercise 4.3.2.

a. Evaluate R–S $\int_{-1}^1 x \, d(|x|)$.

b. Evaluate R–S $\int_0^3 x^2 \, d([x])$.

We now are well positioned to deal with monotonicity of ϕ, the motivation for Stieltjes as he modelled mass distributions.

4.4 Monotonicity of ϕ

We begin with a Fundamental Theorem of Calculus for the R–S integrals.

Theorem 4.4.1 (FTC for Riemann–Stieltjes Integrals). *If f is continuous on the interval $[a, b]$, and ϕ is monotone increasing on $[a, b]$, then R–S $\int_a^b f(x) \, d\phi(x)$ exists.*

Defining a new function F on the interval $[a, b]$ by

$$F(x) = \text{R–S} \int_a^x f(t)\,d\phi(t),$$

then

1. *F is continuous at any point where ϕ is continuous, and*

2. *F is differentiable at each point where ϕ is differentiable (almost everywhere), and at such a point $F' = f\phi'$.*

Proof. We may assume $\phi(b) - \phi(a) > 0$ (recall Exercise 4.1.1). For existence, we may mimic the arguments of Riemann's Integrability Criteria (Section 3.2.1).

We have $\sum_P \inf f \Delta\phi \le \sum_P f \Delta\phi \le \sum_P \sup f \Delta\phi$ since $\Delta\phi$ is nonnegative, and we may conclude that any two Riemann–Stieltjes sums for a partition P are within $\sum_P (\sup f - \inf f)\Delta\phi$ of each other, a quantity that can be made small due to the uniform continuity of the function f.

As for refinements, suppose we add a point $\hat{x}$ with $x_{k-1} < \hat{x} < x_k$. Then

$$\inf_{[x_{k-1}, x_k]} f[\phi(x_k) - \phi(\hat{x}) + \phi(\hat{x}) - \phi(x_{k-1})]$$

$$\le \inf_{[\hat{x}, x_k]} f[\phi(x_k) - \phi(\hat{x})] + \inf_{[x_{k-1}, \hat{x}]} f[\phi(\hat{x}) - \phi(x_{k-1})].$$

As we add points to P, the associated Riemann–Stieltjes sums for the refinements are still within $\sum_P (\sup f - \inf f)\Delta\phi$ of each other. For different partitions, P_1, P_2, $P_1 \cup P_2$ is again a refinement of each and.... Complete the argument for existence.

As for continuity, we may assume $|f| < B$ on the interval $[a, b]$. Thus, recalling Exercise 4.1.1(a), $|F(y) - F(x)| = \left| \text{R–S} \int_x^y f(t)\,d\phi(t) \right| \le B |\phi(y) - \phi(x)|$ implies continuity of F.

Finally, suppose ϕ is differentiable at a point x in the interval (a, b). Because f is continuous on the interval $[a, b]$,

$$\min_{[x, x+h]} f[\phi(x + h) - \phi(x)] \le \text{R–S} \int_x^{x+h} f(t)\,d\phi(t)$$

$$\le \max_{[x, x+h]} f[\phi(x + h) - \phi(x)].$$

Application of the intermediate value theorem to f yields

$$\text{R–S} \int_x^{x+h} f(t)\,d\phi(t) = f(c)[\phi(x + h) - \phi(x)]$$

for h sufficiently small and positive, and c between x and $x + h$. Thus,

$$\left| \frac{F(x+h) - F(x)}{h} - f(x)\phi'(x) \right| = \left| \frac{\text{R-S} \int_x^{x+h} f(t) \, d\phi(t)}{h} - f(x)\phi'(x) \right|$$

$$= \left| f(c)\frac{\phi(x+h) - \phi(x)}{h} - f(x)\phi'(x) \right|$$

$$\leq \left| f(c)\left[\frac{\phi(x+h) - \phi(x)}{h} - \phi'(x) \right] \right|$$

$$+ \left| \phi'(x)[f(x) - f(c)] \right|.$$

The conclusion follows from differentiability of ϕ and continuity of f, at x. Argue when h is small negative. $\square$

Exercise 4.4.1.

a. With C the Cantor function (Section 3.11), calculate

$$\text{R-S} \int_0^1 C(x) \, dC(x) \qquad \text{and} \qquad \text{R-S} \int_0^1 x \, dC(x).$$

b. Assuming that f is continuous and monotone increasing, show that

$$\text{R-S} \int_a^b f(x)df(x) = \frac{1}{2}f^2(b) - \frac{1}{2}f^2(a).$$

4.5   Euler's Summation Formula

A beautiful application of Riemann–Stieltjes integration is given by the Euler Summation Formula.

Theorem 4.5.1 (The Euler Summation Formula). *Suppose f and f' are continuous on the interval $[0, \infty)$. Then*

$$\sum_{k=1}^N f(k) = \text{R} \int_1^N f(x)dx + \frac{f(1) + f(N)}{2} + \text{R} \int_1^N \left(x - [x] - \frac{1}{2} \right) f'(x)dx.$$

As usual, [ ] denotes the greatest integer function.

Proof. The Riemann–Stieltjes integral R–S $\int_1^N f(x)d([x])$ exists, and by Theorem 4.2.1 it is equal to $\sum_{k=2}^N f(k)$, since $[x]$ is a step function on $[1, N]$ with jumps at $2, 3, \ldots, N$.

On the other hand, application of Theorem 4.3.2 tells us that

$$\text{R–S} \int_1^N f(x)\,d([x]) = f(N) \cdot N - f(1) \cdot 1 - \text{R–S} \int_1^N [x]\,df(x)$$

$$= Nf(N) - f(1) - \text{R} \int_1^N [x]f'(x)\,dx$$

by Theorem 4.3.1. Thus,

$$\text{R} \int_1^N \left(x - [x] - \frac{1}{2}\right) f'(x)\,dx$$

$$= \text{R} \int_1^N \left(x - \frac{1}{2}\right) f'(x)\,dx - \text{R} \int_1^N [x]f'(x)\,dx$$

$$= \left(N - \frac{1}{2}\right) f(N) - \frac{1}{2}f(1) - \text{R} \int_1^N f(x)\,dx$$

$$- Nf(N) + f(1) + \text{R–S} \int_1^N f(x)\,d([x])$$

$$= -\frac{1}{2}f(1) - \frac{1}{2}f(N) - \text{R} \int_1^N f(x)\,dx + \sum_{k=1}^N f(k).$$

Solving for $\sum_1^N f(k)$, we have

$$\sum_{k=1}^N f(k) = \text{R} \int_1^N f(x)\,dx + \frac{1}{2}f(1) + \frac{1}{2}f(N)$$

$$+ \text{R} \int_1^N \left(x - [x] - \frac{1}{2}\right) f'(x)\,dx. \quad \square$$

Exercise 4.5.1. Do the following series converge or diverge?

$$\sum_1 \frac{\sin(\sqrt{n})}{n}, \quad \sum_1 \frac{\sin(\ln n)}{n}, \quad \text{and} \quad \sum_1 \frac{\sin(\ln n)}{n^p}, \quad p > 1.$$

Exercise 4.5.2. Let $f(x) = 1/x$ and form the sequence $\{\gamma_n\}$, $\gamma_n = 1 + \frac{1}{2} + \cdots + 1/n - \ln n$. Show $\frac{1}{2} < \gamma_n < 1$. Hint: Tangent lines at $2, 3, \ldots, n$. The limit $\gamma = \lim \gamma_n$ is called Euler's constant.

4.6 Uniform Convergence and R-S Integration

We conclude our treatment of the Riemann–Stieltjes integral with a convergence theorem.

Theorem 4.6.1 (Convergence Theorem for R-S Integrals). *Given that* $\{f_k\}$ *is a sequence of continuous functions converging uniformly to* f *on the interval* $[a, b]$ *and that* ϕ *is monotone increasing on* $[a, b]$. *Then*

1. *the Riemann–Stieltjes integral of* f_k *with respect to* ϕ *exists for all* k,

2. *the Riemann–Stieltjes integral of* f *with respect to* ϕ *exists, and*

3. $\text{R–S} \int_a^b f(x)\, d\phi(x) = \lim \text{R–S} \int_a^b f(x)\, d\phi_k(x)$.

Proof.

$$\left| \text{R–S} \int_a^b f_k(x)\, d\phi(x) - \text{R–S} \int_a^b f(x)\, d\phi(x) \right|$$

$$\leq \text{R–S} \int_a^b |f_k(x) - f(x)|\, d\phi(x). \qquad \square$$

Exercise 4.6.1. Calculate $\lim \text{R–S} \int_0^{\pi/2} \left(1 - (x/n)\right)^n d\left(\sin(x)\right)$.

4.7   References

1. Apostol, Tom. *Mathematical Analysis*. 2nd edition. Reading, Mass.: Addison-Wesley, 1974.

2. Birkhoff, Garrett. *A Source Book in Classical Analysis*. Boston: Harvard University Press, 1973.

CHAPTER **5**

Lebesgue Measure

The scientist does not study nature because it is useful; he studies it because he delights in it, and he delights in it because it is beautiful. If nature were not beautiful, it would not be worth knowing, and if nature were not worth knowing, life would not be worth living. Of course I do not here speak of that beauty that strikes the senses, the beauty of qualities and appearances, not that I undervalue such beauty, far from it, but it has nothing to do with science; I mean that profounder beauty which comes from the harmonious order of the parts, and which a pure intelligence can grasp.
            — Henri Poincaré

Vito Volterra's 1881 example of a function having a bounded derivative that was not Riemann integrable (Section 3.12) prompted Henri Lebesgue to develop a method of integration to remedy this shortcoming: The Lebesgue integral of a bounded derivative returns the original function.

5.1 Lebesgue's Idea

Lebesgue's integration process was fundamentally different from that of his predecessors. His simple but brilliant idea: *Partition the range of the function rather than its domain.*

How does this work? Suppose the function f is bounded on the interval $[a, b]$, say $\alpha < f < \beta$. Partition the interval $[\alpha, \beta]$,

$$\alpha = y_0 < y_1 < y_2 < \cdots < y_{n-1} < y_n = \beta,$$

with E_k denoting the points of the interval $[a, b]$ for which $y_{k-1} \leq f < y_k$. We have partitioned the interval $[a, b]$ into disjoint sets E_k with $[a, b] = \cup E_k$.

In each nonempty set E_k, pick a point c_k and form the sum $\sum f(c_k) \cdot$ (length of E_k). Observe that

$$\sum y_{k-1} \cdot (\text{length of } E_k) \leq \sum f(c_k) \cdot (\text{length of } E_k)$$
$$\leq \sum y_k \cdot (\text{length of } E_k),$$

and that the absolute value of the difference of any two such sums,

$$\left| \sum f(c_k) \cdot (\text{length of } E_k) - \sum f(\hat{c}_k) \cdot (\text{length of } E_k) \right|$$

is bounded by $\max_{k=1,\ldots,n} (y_k - y_{k-1}) \cdot (b - a)$. Apparently...

What do we mean by the *length* of E_k? Certainly if E_k is an interval, there's no problem. But suppose, for example, the function f under discussion is the absolute value function on the interval $[-1, 1]$. Well, E_k could be $[y_{k-1}, y_k) \cup (-y_k, -y_{k-1}]$. No problem, then: the length of E_k is $2(y_k - y_{k-1})$.

But then we have Dirichlet's function on the interval $[0, 1]$, one on the rationals and zero on the irrationals. We would have to calculate the length of the rational numbers in $[0, 1]$ and the length of the irrational numbers in $[0,1]$.

The sets E_k may be very complicated. This integral, if it is to have any power, requires us to assign a length, to assign a *measure*, to a large variety of sets.

5.2   Measurable Sets

Let's think about what to put on a Measuring Wish List...

1. We want to measure all sets of real numbers.

2. The measure of a subset must not exceed the measure of the set.

3. A point should have measure zero.

4. The measure of an interval should be its length.

5. Translation of a set of real numbers should not alter its measure.

6. The measure of the whole must be the sum of the measures of its parts.

Unfortunately, we cannot define a measure that meets all these requirements. For one thing, Wishes 3 and 6 would imply that the measure of any set of real numbers is zero.

Yet linearity of the integral demands Wish 6, and Wish 6 seems so reasonable (with appropriate modification) that we will back off on the first item on our list. We will keep additivity at the expense of measuring not all — but as it turns out, enough — sets of real numbers. The idea here is that we will define a measure, commonly referred to as the *Lebesgue outer measure*, that measures all sets of real numbers (not always in an additive fashion); we will discard those sets where additivity fails.

Recall that every set of real numbers may be covered by a countable collection of open intervals. In fact, a set has many such coverings, whose lengths are easily calculated: the length of a cover will be the sum of the lengths of the open intervals making up the cover. The infimum of this set of extended real numbers is called the *Lebesgue outer measure of the set*. Formally, for any set A of real numbers, the Lebesgue outer measure of A, written $\mu^*(A)$, is given by

$$\mu^*(A) = \inf\left\{\sum \ell(I_k) \mid A \subset \cup I_k, \ I_k \text{ open intervals}\right\}.$$

5.2.1 The Wish List and Lebesgue Outer Measure

What do we have if we adapt our Wish List to the Lebesgue outer measure?

1. Every set of real numbers has a Lebesgue outer measure.

2. Since any cover of a set will be a cover of a subset, the infimum property yields monotonicity.

3. The empty set is a subset of every set; $\{a\} \subset (a - \epsilon, a + \epsilon)$; and the Lebesgue outer measure of the empty set, or a singleton set, is zero.

4. Because (a, b) is a cover for (a, b), $\mu^*\big((a, b)\big) \leq b - a$. (In fact, generally the outer measure of an interval is its length.)

5. Translation invariance is not a problem.

6. Additivity remains.

The last three points require some explanation.

Adapted Wish 4. We will show that $\mu^*\big((a, b)\big) = b - a$. If $\cup I_k$ is a cover of (a, b), then $\cup I_k$, with $(a - \epsilon, a + \epsilon)$ and $(b - \epsilon, b + \epsilon)$, is a cover of $[a, b]$. We have an open cover of a compact set. By the Heine–Borel

Theorem, $a_{n_1} < a < b_{n_1}, a_{n_2} < b_{n_1} < b_{n_2}, \ldots, a_{n_k} < b < b_{n_k}$, and

$$
\begin{aligned}
b - a &= (b - a_{n_k}) + (a_{n_k} - a_{n_{k-1}}) + \cdots + (a_{n_2} - a) \\
&\le (b_{n_k} - a_{n_k}) + (b_{n_{k-1}} - a_{n_{k-1}}) + \cdots + (b_{n_1} - a_{n_1}) \\
&\le \sum \ell(I_k) + 4\epsilon.
\end{aligned}
$$

That is, $b-a$ is a lower bound for the length of any cover of (a, b), and since the outer measure of (a, b) is the largest lower bound, $\mu^*((a, b)) \ge b - a$.

Thus, the outer measure of the interval (a, b) is simply its length.

For unbounded intervals, say (a, ∞), the open interval (a, c) is a subset of (a, ∞) for every real number c larger than a, and monotonicity completes the argument. Show $\mu^*(I) = \ell(I)$ for other kinds of intervals, such as $[a, b)$ and $(-\infty, b]$.

Adapted Wish 5. Why is translation invariance not a problem? Because the length of an interval does not change under translation. If $A \subset \cup I_k$, then $c + A \subset \cup(c + I_k)$ and

$$
\mu^*(c + A) \le \sum \ell(c + I_k) = \sum \ell(I_k).
$$

That is, $\mu^*(c + A)$ is a lower bound for the length of any cover of A. In other words, $\mu^*(c + A) \le \mu^*(A)$ by the infimum property.

Starting with a cover of $c + A$, $\cup I_k$, argue the other direction.

Adapted Wish 6. Additivity remains. But is the Lebesgue outer measure additive? Is the Lebesgue outer measure of a collection of disjoint sets the sum of the Lebesgue outer measures of the individual sets? If so, our quest would be over.

In 1905, Vitali constructed a Lebesgue nonmeasurable set of real numbers; additivity does not always hold. What we can show, though, is that the Lebesgue outer measure of any countable collection of sets of real numbers (not necessarily disjoint) is at most the sum of Lebesgue outer measures of the individual sets:

$$
\mu^*(\cup A_k) \le \sum \mu^*(A_k).
$$

If $\mu^*(A_k) = \infty$ for any k, we are done. Assume $\mu^*(A_k) < \infty$ for all k. We have an open cover $\cup_n I_{kn}$ of each A_k with

$$
\mu^*(A_k) \le \sum_n \ell(I_{kn}) < \mu^*(A_k) + \epsilon/2^k, \quad \text{for } \epsilon > 0. \quad \text{(infimum property)}
$$

The collection of open intervals I_{kn}, for $k, n = 1, 2, \ldots$, is a cover of $\cup A_k$, and

$$\mu^*(\cup A_k) \leq \sum_k \sum_n \ell(I_{kn}) < \sum_k \mu^*(A_k) + \epsilon.$$

Thus

$$\mu^*(\cup A_k) \leq \sum \mu^*(A_k). \qquad \text{(subadditivity)}$$

Vitali's example is particularly upsetting. We are so close with this Lebesgue outer measure. What to do? We need an ingenious way around this.

5.3   Lebesgue Measurable Sets and Carathéodory

Here is an ingenious idea: Any set A of real numbers can be decomposed, relative to a set E, into two disjoint sets, its intersections with E and the complement of E, $A \cap E$ and $A \cap E^c$. Additivity would suggest that $\mu^*(A) = \mu^*(A \cap E) + \mu^*(A \cap E^c)$.

Let us then select from the collection of all subsets of real numbers precisely those sets E that interact in this fashion. Such sets E will be called *Lebesgue measurable sets*.

Definition 5.3.1 (Carathéodory's Measurability Criterion, 1914). A set of real numbers E is Lebesgue measurable if

$$\mu^*(A) = \mu^*(A \cap E) + \mu^*(A \cap E^c)$$

holds for every set of real numbers A.

So we have a criterion for measurability: Select a set E, and check whether it "splits" every set A of real numbers in an additive fashion. If the answer is yes, keep E. Otherwise, discard E. This seems straightforward enough. But we may end up discarding most of our collection of subsets of the real numbers.

Exercise 5.3.1. Show that the empty set and the set of all real numbers satisfy Carathéodory's measurability criterion. If E is Lebesgue measurable, show that the complement of E is Lebesgue measurable.

5.3.1   Intervals

Are intervals Lebesgue measurable? We know that the Lebesgue outer measure of an interval is its length. We want to show that intervals satisfy

Carathéodory's measurability criterion. Of course, for an interval that can be written as the union of two disjoint intervals, length is additive; as it turns out, this observation will yield measurability of intervals.

Because $(a, b) = (a, \infty) \cap (-\infty, b)$ when $a < b$ — and because we have to start somewhere — we will show that (a, ∞) is Lebesgue measurable. Actually, we need only show

$$\mu^*(A) \geq \mu^*(A \cap (a, \infty)) + \mu^*(A \cap (-\infty, a]),$$

since the reverse inequality holds by the subadditivity of the Lebesgue outer measure. We may also assume $\mu^*(A) < \infty$. By the infimum property, we may select a"tight" open cover $\cup I_k$ of A; that is, $\mu^*(A) \leq \sum \ell(I_k) < \mu^*(A) + \epsilon$, for $\epsilon > 0$. Thus

$$\mu^*(A \cap (a, \infty)) + \mu^*(A \cap (-\infty, a])$$
$$\leq \mu^*((\cup I_k) \cap (a, \infty)) + \mu^*((\cup I_k) \cap (-\infty, a]) \quad \text{(monotonicity)}$$
$$\leq \sum \mu^*(I_k \cap (a, \infty)) + \sum \mu^*(I_k \cap (-\infty, a]) \quad \text{(subadditivity)}$$
$$= \sum [\ell(I_k \cap (a, \infty)) + \ell(I_k \cap (-\infty, a])] = \sum \ell(I_k)$$
$$< \mu^*(A) + \epsilon,$$

and we have

$$\mu^*(A \cap (a, \infty)) + \mu^*(A \cap (-\infty, a]) \leq \mu^*(A).$$

This is what we wanted to show.

Demonstrate the Lebesgue measurability of $(-\infty, b)$, (a, b), and so on.

Intervals are Lebesgue measurable. We have devised a complicated measure, *an infinitum of infinite sums*, to measure something we already knew the measure of — the length of an interval. Isn't mathematics humbling?

We know a few Lebesgue measurable sets — the empty set, its complement, and intervals and their complements. How about unions and intersections of measurable sets? Most important of all, if $\{E_k\}$ is a sequence of mutually disjoint Lebesgue measurable sets, is $\cup E_k$ Lebesgue measurable? Furthermore, does additivity hold: $\mu^*(\cup E_k) = \sum \mu^*(E_k)$? Is the Lebesgue outer measure, when restricted to Lebesgue measurable sets, countably additive?

It is time to discuss *sigma algebras* and investigate why they are important in the context of measure theory.

5.4 Sigma Algebras

Definition 5.4.1 (A Sigma Algebra). Given a space Ω, a collection $\mathcal{O}$ of subsets of Ω is said to be a sigma algebra provided the following properties hold:

1. The empty set belongs to this collection.

2. If a set is in this collection, its complement is in the collection.

3. Given any sequence of sets in this collection, their union is in the collection.

Carathéodory showed that the collection of subsets of real numbers that satisfy his measurability criterion forms a sigma algebra and, moreover, that the Lebesgue outer measure is countably additive on this sigma algebra. This is the essence of the next theorem, one of the two most important results regarding Lebesgue measure. The other crucial result is Theorem 5.5.1.

Theorem 5.4.1 (Carathéodory, 1914). *Define the Lebesgue outer measure μ^* of any set of real numbers E as follows:*

$$\mu^*(E) = \inf\left\{\sum \ell(I_k) \mid E \subset \cup I_k,\ I_k\ \text{open intervals}\right\},$$

where $\ell(I_k) = b_k - a_k$, a_k, b_k are extended real numbers, $a_k < b_k$. Then

1. *The collection of sets E of real numbers that satisfy Carathéodory's measurability criterion, $\mu^*(A) = \mu^*(A \cap E) + \mu^*(A \cap E^c)$ for every set of real numbers A, forms a sigma algebra $\mathcal{M}$.*

2. *The Lebesgue outer measure is countably additive on $\mathcal{M}$: For any sequence $\{E_k\}$ of mutually disjoint Lebesgue measurable sets,*

$$\mu^*\left(\cup_1^\infty E_k\right) = \sum_1^\infty \mu^*(E_k).$$

Proof. Show the first two requirements for a collection of sets to form a sigma algebra are met. The third requirement, that $\cup E_k$ be a member of $\mathcal{M}$, may be demonstrated as follows.

Start with an easier problem: The intersection of two measurable sets, and the union of two measurable sets, is measurable. Show that

$$A \cap (E_1 \cup E_2) = (A \cap E_1) \cup (A \cap E_1^c \cap E_2) \quad \text{and}$$
$$A \cap (E_1^c \cap E_2^c) = A \cap (E_1 \cup E_2)^c.$$

Then

$$\mu^*(A) = \mu^*(A \cap E_1) + \mu^*(A \cap E_1^c) \quad \text{measurability of } E_1$$
$$= \mu^*(A \cap E_1) + \mu^*\big((A \cap E_1^c) \cap E_2\big) + \mu^*\big((A \cap E_1^c) \cap E_2^c\big)$$

measurability of E_2

$$\geq \mu^*\big((A \cap E_1) \cup (A \cap E_1^c) \cap E_2\big) + \mu^*\big((A \cap E_1^c) \cap E_2^c\big)$$

subadditivity

$$= \mu^*\big(A \cap (E_1 \cup E_2)\big) + \mu^*\big(A \cap (E_1 \cup E_2)^c\big) \quad \text{subadditivity}$$
$$\geq \mu^*(A).$$

The union of two Lebesgue measurable sets is Lebesgue measurable. Complementation shows the intersection of two measurable sets is measurable: $(E_1 \cap E_2) = (E_1^c \cup E_2^c)^c$. Measurability of finite unions and finite intersections follows by induction.

As for $(\cup_1^\infty E_k)$, we may assume the E_k are mutually disjoint. In this case,

$$\mu^*(E_1 \cup E_2) = \mu^*\big((E_1 \cup E_2) \cap E_2\big) + \mu^*\big((E_1 \cup E_2) \cap E_2^c\big)$$
$$= \mu^*(E_2) + \mu^*(E_1),$$

since E_1 and E_2 are disjoint. By induction, we have finite additivity and

$$\mu^*\Big(A \cap (U_1^n E_k)\Big) = \mu^*\Big(A \cap (U_1^n E_k) \cap E_K\Big) + \mu^*\Big(A \cap (U_1^n E_k) \cap E_n^c\Big)$$

$$= \mu^*(A \cap E_n) + \sum_1^{n-1} n^*(A \cap E_k)$$

$$= \sum_1^n \mu^*(A \cap E_k).$$

Since

$$\mu^*(A) = \mu^*\Big(A \cap (U_1^n E_k)\Big) + \mu^*\Big(A \cap (U_1^n E_k)^c\Big)$$

$$\geq \sum_1^n \mu^*(A \cap E_k) + \mu^*\Big(A \cap (U_1^\infty E_k)^c\Big),$$

independent of n, we have

$$\mu^*(A) \geq \sum_1^\infty \mu^*(A \cap E_k) + \mu^*\Big(A \cap (U_1^\infty E_k)^c\Big)$$

$$\geq \mu^*\Big(A \cap (U_1^\infty E_k)\Big) + \mu^*\Big(A \cap (U_1^\infty E_k)^c\Big),$$

and $\cup_1^\infty E_k$ is a member of $\mathcal{M}$, $\cup_1^\infty E_k$ is Lebesgue measurable.

We have shown that the collection of sets of real numbers satisfying Carathéodory's measurability criterion, $\mathcal{M}$, is a sigma algebra. The task of showing countable additivity remains.

We have $\mu^*(\cup E_k) \leq \sum \mu^*(E_k)$ by the subadditivity of Lebesgue outer measure. For the other direction, we have finite additivity,

$$\sum_1^n \mu^*(E_k) = \mu^*(\cup_1^n E_k) \leq \mu^* \left(\cup_1^\infty E_k \right) \leq \sum_1^\infty \mu^*(E_k),$$

independent of n, and hence

$$\sum_1^\infty \mu^*(E_k) \leq \mu^* \left(\cup_1^\infty E_k \right) \leq \sum_1^\infty \mu^*(E_k).$$

This completes the proof. $\square$

5.5   Borel Sets

The Lebesgue outer measure is countably additive when restricted to the sigma algebra of sets $\mathcal{M}$ satisfying Carathéodory's measurability criterion. For $E \in \mathcal{M}$, we will write $\mu(E)$ for $\mu^*(E)$. It is to be understood that writing μ assumes the set is Lebesgue measurable. *This measure μ may be the measure needed for the Lebesgue integral.*

Recall that the only sets that we know are measurable (i.e., are in $\mathcal{M}$) are ϕ, R, intervals, and their complements.

Not quite. We have just shown that we can measure countable unions (rational numbers, for example), their complements (irrational numbers), and so on — some very complicated sets. It is time to introduce *Borel sets*.

We have shown that open intervals are Lebesgue measurable; open intervals are in the sigma algebra $\mathcal{M}$. Show that the intersection of all sigma algebras that contain a given collection of sets of real numbers is again a sigma algebra. This sigma algebra is said to be *generated* by the given collection. We give a special name, the *Borel sigma algebra*, to that sigma algebra generated by the collection of open intervals, and we write $\mathcal{B}$.

Since we have shown that open intervals are Lebesgue measurable, evidently $\mathcal{B} \subseteq \mathcal{M}$. We have, almost unexpectedly, arrived at a very important result.

Theorem 5.5.1. *Every Borel set of real numbers is Lebesgue measurable.*

We can measure the Borel sets (we are hard-pressed to find non-Borel Lebesgue measurable sets), and measuring works as expected for measurable sets: When the whole is decomposed into a countable number of disjoint measurable parts, the measure of the whole is the sum of the measures of its parts. Here are some exercises in measuring.

Exercise 5.5.1. Show that the following assertions are true.

a. $\mu(I) = \ell(I)$. Hint: $(a, b) \subseteq (a, b)$. So $\mu^*((a, b)) \leq b - a$, and Section 5.2.1.

b. $\mu((0, 1]) = \sum_{1}^{\infty} \mu\left(\left(\frac{1}{k+1}, \frac{1}{k}\right]\right)$.

c. Singleton sets are measurable and have measure zero. Hint: $\{a\} = \cap (a - 1/k, a + 1/k)$.

d. Any countable set is measurable and has measure zero. Hint: $\{a_1, a_2, \ldots, a_k, \ldots\} = \cup a_k$, and $\mu(a_k) < \mu\left(a_k - \frac{\epsilon}{2^k}, a_k + \frac{\epsilon}{2^k}\right)$.

e. The rationals are measurable and have measure zero.

f. The irrational numbers in $[0, 1]$ are measurable. What is their measure? (A set of positive measure containing no intervals.)

g. Any set with $\mu^*(E) = 0$ is Lebesgue measurable. Hint: $\mu^*(A) \leq \mu^*(A \cap E) + \mu^*(A \cap E^c) \leq \mu^*(E) + \mu^*(A) = \mu^*(A)$.

In fact, there are uncountable sets of Lebesgue measure zero.

Exercise 5.5.2. Show that the measure of the Cantor set is zero.

5.6 Approximating Measurable Sets

We are familiar with the ideas of open sets, closed sets, compact sets, and so on. Do we have relationships between these topological notions and measurability of sets? It turns out that Lebesgue measurable sets may be closely approximated by open sets "from the outside" and closed sets "from the inside."

Theorem 5.6.1. *For any set E of real numbers the following statements are equivalent:*

1. *E is Lebesgue measurable in the sense of Carathéodory.*

2. *Given $\epsilon > 0$, we have an open set G containing E so that $\mu^*(G-E) < \epsilon$.*

3. *Given $\epsilon > 0$, we have a closed set F contained in E so that $\mu^*(E - F) < \epsilon$.*

Proof. Assume E is measurable. We will show that the measure of E may be approximated by the measure of an open set. By the infimum property, we have an open cover $\cup I_k$ so that $\mu(E) \le \mu(\cup I_k) < \mu(E) + \epsilon$; and because E is measurable,

$$\mu(\cup I_k) = \mu\big((\cup I_k) \cap E\big) + \mu\big((\cup I_k) - E\big) = \mu(E) + \mu\big((\cup I_k) - E\big).$$

If the measure of E is finite, subtraction yields the desired result. Otherwise, let $E_n = E \cap [-n, n]$ and argue as before. So statement 1 implies statement 2.

Show, using complementation, that open set approximation yields closed set approximation. So statement 2 implies statement 3.

As for the third conclusion, suppose we have a closed subset F of E with $\mu^*(E - F) < \epsilon$. It is sufficient to show $\mu^*(A) \ge \mu^*(A \cap E) + \mu^*(A \cap E^c)$ with $\mu^*(A)$ being finite. We have

$$\begin{aligned}
\mu^*(A \cap E) &= \mu^*\Big(A \cap \big((E - F) \cup F\big)\Big) \\
&\le \mu^*\big(A \cap (E - F)\big) + \mu^*(A \cap F) \\
&< \epsilon + \mu^*(A \cap F)
\end{aligned}$$

and

$$\begin{aligned}
\mu^*(A \cap E^c) &= \mu^*\Big(A \cap \big((E - F) \cup F\big)^c\Big) \\
&= \mu^*\big(A \cap (E - F)^c \cap F^c\big) \\
&\le \mu^*(A \cap F^c).
\end{aligned}$$

Because closed sets are measurable, $\mu^*(A \cap F) + \mu^*(A \cap F^c) = \mu^*(A)$, so addition yields

$$\mu^*(A \cap E) + \mu^*(A \cap E^c) \le \epsilon + \mu^*(A)$$

and we have the desired conclusion. Thus statement 3 implies statement 1, and the equivalence of 1, 2, and 3 follows. $\square$

We return to Borel sets and some exercises "completing" $\mathcal{B}$.

Exercise 5.6.1.

   a. Show that for any measurable set E, we have a Borel set B_1 so that E is a subset of B_1 and $\mu(B_1 - E) = 0$. Hint: From the previous result we have an open set G_n so that $\mu(G_n - E) < 1/n$; $B_1 = \cap G_n$.

   b. Show that for any measurable set E we have a Borel set B_2 so that B_2 is a subset of E and $\mu(E - B_2) = 0$. Hint: $B_2 = \cup F_n$.

   c. Show that every Lebesgue measurable set of real numbers is the union of a Borel set and a set of Lebesgue measure zero. Hint: $E = B_2 \cup (E - B_2)$.

Adjoining the sets of Lebesgue measure zero to the Borel sets creates the Lebesgue measurable sets.

5.6.1 Vitali's Covering Theorem

Here is a result by Giuseppe Vitali (1875–1932) that we will find very useful. We begin with a definition.

Definition 5.6.1 (Vitali Cover). Let A be a nonempty set of real numbers. A Vitali cover of A is a collection of closed intervals of arbitrarily small length that cover each point of A.

Thus every point of A is in an arbitrarily small closed interval from this collection.

Theorem 5.6.2 (Vitali, 1908). *Given a set of real numbers A whose Lebesgue outer measure is finite and a Vitali cover of this set A. Then given an $\epsilon > 0$, there exists a finite, disjoint collection of closed intervals $I_1, I_2, \ldots, I_n$ from this Vitali cover of A so that*

$$\mu^*(A - \cup_{k=1}^n I_k) < \epsilon.$$

This finite collection can be extended to a countable collection of mutually disjoint, closed intervals $I_1, I_2, \ldots, I_n, \ldots$ from this Vitali cover of A so that $\mu^*(A - \cup_1^\infty I_k) = 0$.

An outline of the argument for this result may be found in the wonderful book by Stromberg (1981).

5.7 Measurable Functions

Recall that Cauchy's integration process was very effective for continuous functions, functions that preserve openness under inverse images. Since every open set of real numbers is the union of a countable collection of disjoint open intervals, and since inverse images behave nicely — $f^{-1}(\cup I_k) = \cup f^{-1}(I_k)$ — why not try measurement of inverse images of intervals? This was Lebesgue's idea (Section 5.1).

Definition 5.7.1 (Lebesgue Measurable Functions). A real-valued function f that is defined on a measurable set E is said to be Lebesgue measurable on E if the inverse images under f of intervals of real numbers are measurable subsets of E.

How would we check to see if a function is Lebesgue measurable? Do we have to argue all types of intervals?

Exercise 5.7.1.

a. Assume $f^{-1}((c, \infty))$ is Lebesgue measurable for every real number c. Show that the inverse images of $[a, \infty)$, $(-\infty, a)$, $(-\infty, a]$, (a, b), $(a, b]$, $[a, b)$, $[a, b]$, are Lebesgue measurable. Hint: $[a, \infty) = \cap(a - (1/k), \infty)$ and $f^{-1}([a, \infty)) = \cap f^{-1}(a - (1/k), \infty)$.

b. Show that the following function is Lebesgue measurable on $[-1, 2]$:

$$f(x) = \begin{cases} x^2 & x < 1, \\ 2 & x = 1, \\ 2 - x & x > 1. \end{cases}$$

Just what kind of functions are Lebesgue measurable?

5.7.1 Continuous Functions Defined on Measurable Sets

We will show that continuous functions defined on measurable sets are Lebesgue measurable. Consider the equation $A = f^{-1}((c, \infty)) = \{x \in E \mid f(x) > c\}$.

If A is empty, we are done. Otherwise, for each x in A, we have $\delta(x) > 0$ so that for z belonging to the interval $(x - \delta(x), x + \delta(x))$, $f(z) > c$:

$$A = \cup_{x \in A}\left((x - \delta(x), x + \delta(x)) \cap E\right) = \left(\cup_{x \in A}(x - \delta(x), x + \delta(x))\right) \cap E.$$

5.7.2 Riemann Integrable Functions

Cauchy integrable functions are Lebesgue measurable functions. How about Riemann integrable functions? We will show that Riemann integrable functions are Lebesgue measurable functions.

As a first step, we will suppose f and g are defined on a measurable set E, that f is Lebesgue measurable on E, and that $g = f$ except on a subset Z of E of Lebesgue measure zero (almost everywhere). We claim that g is Lebesgue measurable on E.

Consider the following relationships:

$$\{x \in E \mid g(x) > c\} = \{x \in Z \mid f(x) > g(x) > c\}$$
$$\cup \{x \in E - Z \mid f(x) = g(x) > c\}$$
$$\cup \{x \in Z \mid g(x) > f(x) > c\}$$
$$\cup \{x \in Z \mid g(x) > c = f(x)\}$$
$$\cup \{x \in Z \mid g(x) > c > f(x)\}$$
$$= \{x \in E \mid f(x) > c\} \cup \{x \in Z \mid g(x) > c\}$$
$$\cup \{x \in Z \mid g(x) > c > f(x)\}.$$

If f is Riemann integrable in $[a, b]$, then f's discontinuities form a set of measure zero, f is continuous on $[a, b] - Z$, and f is measurable on $[a, b] - Z$.

Let $g = f$ on $[a, b] - Z$ and 0 on A. Then $g = f$ almost everywhere, g is measurable on $[a, b]$, and thus f is measurable on $[a, b]$.

Lebesgue measurable functions may not be Riemann integrable.

Exercise 5.7.2. Let

$$f(x) = \begin{cases} 1 & x \text{ rational}, \\ 0 & x \text{ irrational}, \end{cases} \qquad 0 \leq x \leq 1.$$

Show that f is Lebesgue measurable on $[0, 1]$.

5.7.3 Limiting Operations and Measurability

An important property of sequences of measurable functions is that measurability is preserved under many limiting operations.

Theorem 5.7.1 (Limiting Operations and Measurability). *If $\{f_k\}$ is a sequence of Lebesgue measurable functions defined on a measurable set E with $\lim f_k = f$ pointwise on E, then f is Lebesgue measurable on E.*

For example, if

$$g_k(x) \equiv \sup\{f_k(x), f_{k+1}(x), \ldots\} = \cup_{n \geq k}\{x \in E \,|\, f_n(x) > c\},$$

then g_k is measurable.

If $h_k(x) \equiv \inf\{f_k(x), f_{k+1}(x), \ldots\} = \cup_{n \geq k}\{x \in E \,|\, f_n(x) < c\}$, then h_k is measurable.

If $\limsup f_k = \lim g_k$, then $\{x \in E \,|\, \limsup f_k(x) < c\} = \cup_{k=1}^{\infty}\{x \in E \,|\, g_k(x) < c\}$, and so on.

However, Riemann integrable functions do not share this property. Limiting operations may not preserve Riemann integrability. For example, let $r_1, r_2, \ldots$ be any enumeration of the rational numbers in $(0, 1)$, and define a sequence $\{f_k\}$ of Riemann integrable functions by

$$f_k(x) = \begin{cases} 1 & x = r_1, r_2, \ldots, r_k, \\ 0 & \text{otherwise.} \end{cases}$$

Then $R \int_0^1 f_k(x)\, dx = 0$, but the pointwise limit of the sequence $\{f_k\}$ is the Dirichlet function, which is not Riemann integrable.

5.7.4 Simple Functions

Simple functions have a finite number of values. That is, for a measurable set E, a simple function ϕ can be written as a finite linear combination of characteristic functions:

$$\phi(x) = \sum_{k=1}^{n} c_k \chi_{E_k}(x)$$

where c_k are real numbers, E_k are mutually disjoint measurable sets with $E = \cup_{k=1}^{n} E_k$, and $\chi_{E_k}(x)$ is the characteristic function

$$\chi_{E_k}(x) = \begin{cases} 1 & x \in E_k, \\ 0 & x \notin E_k. \end{cases}$$

The reader may show that simple functions are measurable functions.

Theorem 5.7.2 (Approximating Measurable Functions by Simple Functions). *For any measurable function f defined on a measurable set E, there exists a sequence of simple functions $\{\phi_k\}$ on E so that $\lim \phi_k = f$ for all E.*

If f is bounded on E, then $\lim \phi_k = f$ uniformly on E. If f is nonnegative, the sequence $\{\phi_k\}$ may be constructed so that it is a monotonically increasing sequence.

Proof. We follow Lebesgue's idea: partition the range of f. We may assume that f is nonnegative; that is, we have

$$f = \frac{|f| + f}{2} - \frac{|f| - f}{2}$$

for $\frac{|f|+f}{2}$, $\frac{|f|-f}{2}$ measurable and nonnegative. Let $[0, \infty) = [0, 1) \cup [1, 2) \cup \cdots \cup [n-1, n) \cup [n, \infty)$, and partition $[0, \infty)$ into $2^n + 2^n + \cdots + 2^n + 1 = n2^n + 1$ disjoint subintervals.

Define ϕ_n by

$$\phi_n(x) = \sum_{k=1}^{n2^n} \frac{k-1}{2^n} \chi_{E_{nk}}(x) + n \chi_{F_n}(x)$$

$$\text{with } E_{nk} = \left\{ x \in E \;\middle|\; \frac{k-1}{2^n} \le f(x) < \frac{k}{2^n} \right\}, \quad F_n = [n, \infty).$$

Note that $E_{nk} = E_{n+1\,2k-1} \cup E_{n+1\,2k}$. The reader may complete the argument. $\square$

5.7.5 Pointwise Convergence Is Almost Uniform Convergence

Because uniform convergence transfers many nice properties to the limit function, we look for conditions that generate uniform convergence. For sequences of measurable functions, we have a remarkable theorem showing that pointwise convergence is almost uniform convergence. The theorem is due to Dimitri Egoroff.

Theorem 5.7.3 (Egoroff, 1911). *Suppose $\{f_k\}$ is a sequence of measurable functions that converges to a real-valued function f almost everywhere on the interval $[a, b]$. Then for any $\delta > 0$, we have a measurable subset E of $[a, b]$ so that $\mu(E) < \delta$ and the sequence $\{f_k\}$ converges uniformly to f on $[a, b] - E$.*

5.8 More Measureable Functions

In addition to continuous functions, differentiable functions, monotone functions, and Riemann integrable functions, two other classes of measurable functions — functions of bounded variation and absolutely continuous functions — will be of interest as we develop the Fundamental Theorems of Calculus for Lebesgue integrals.

5.8.1   Functions of Bounded Variation

Camille Jordan (1838–1922) offered the following definition in 1881.

Definition 5.8.1 (Bounded Variation). A function is said to be of bounded variation on an interval $[a, b]$ provided $\sum_P |f(x_k) - f(x_{k-1})|$ is bounded for all possible partitions P of $[a, b]$, $a = x_0 < x_1 < \cdots < x_n = b$.

Exercise 5.8.1.

a. Show that continuous functions may not be of bounded variation, for example

$$f(x) = \begin{cases} x \sin(\pi/x) & 0 < x \le 1, \\ 0 & 0 = x. \end{cases}$$

Hint: $x_k = 2/(2k + 1)$, $k = 1, 2, \ldots$.

b. Show that monotonic functions are of bounded variation.
Hint: $\sum_P |f(x_k) - f(x_{k-1})| \le |f(b) - f(a)|$.

5.8.2   Functions of Bounded Variation and Monotone Functions

Theorem 5.8.1 (Jordan, 1894). *Functions of bounded variation are the difference of two monotone increasing functions.*

Clearly the difference of two monotone functions is a function of bounded variation. Now suppose f is a function of bounded variation on $[a, b]$. Define a new function V, the variation of f on $[a, b]$, by

$$V(x) \equiv \sup_P \sum_P |f(x_k) - f(x_{k-1})|$$

over all partitions P of $[a, x]$, with $a \le x \le b$. The function V is certainly monotone increasing. Since $f = V - (V - f)$, we need only show that $V - f$ is monotone increasing, that is, for $x < y$,

$$V(x) - f(x) \le V(y) - f(y) \qquad \text{or} \qquad f(y) - f(x) \le V(y) - V(x).$$

But the variation on $[x, y]$ is at least as large as $|f(x) - f(y)|$ (trivial partition).

It will be shown (Theorem 5.10.1) that monotone functions are differentiable almost everywhere. Thus functions of bounded variation are differentiable almost everywhere.

This seems simple enough, the difference of two monotonic functions characterizing functions of bounded variations. What else is known? Kannan and Krueger (1996) offered the following observation.

Example 5.8.1. A function of bounded variation is the difference of two monotone functions, but it need not be monotonic on any subinterval of its domain.

This needs a closer look.

Let $r_1, r_2, \ldots$ be an enumeration of the rational numbers in $(0, 1)$, and let $0 < \alpha < 1$. Define f on $[0, 1]$ by

$$f(x) = \begin{cases} \alpha^k & x = r_k, k = 1, 2, \ldots, \\ 0 & x \neq r_k. \end{cases}$$

We claim that the total variation of f on the interval $[0, 1]$ is $2\alpha/(1 - \alpha)$. Construct a partition $P = \{0, x_1, x_2, \ldots, x_{2n-1}, 1\}$, with the "odds" being $\{x_1, x_3, \ldots, x_{2n-1}\} = \{r_1, r_2, \ldots, r_n\}$ and the "evens" irrational numbers in $[0, 1]$. Then

$$\sum_P |f(x_k) - f(x_{k-1})| = 2(\alpha^1 + \alpha^2 + \cdots + \alpha^n) \qquad \text{and}$$

$$V(1) \geq 2 \sum_1^n \alpha^k, \qquad n = 1, 2, \ldots.$$

Thus $V(1) \geq 2\alpha/(1 - \alpha)$.

On the other hand, for any partition $P = \{0, x_1, x_2, \ldots, x_{n-1}, 1\}$, r_1 will be in one of these subintervals or will be an endpoint. Regardless, r_1 will be in exactly one of the (x_{k-1}, x_{k+1}), with $k = 1, 2, \ldots, n - 1$, and r_1 makes a contribution to the variation only if it is an endpoint, an evaluation point of f. Similarly for $r_2, r_3, \ldots, r_{n-1}$.

So the worst case occurs when the $n - 1$ rational numbers $r_1, r_2, \ldots, r_{n-1}$ occur as partition points. Thus the variation for any partition is bounded by

$$2 \left(\alpha^1 + \alpha^2 + \cdots + \alpha^{n-1} \right) < \frac{2\alpha}{1 - \alpha}.$$

5.8.3 Absolutely Continuous Functions

Vito Vitali developed this definition in 1904.

Definition 5.8.2 (Absolutely Continuous Function). A function f on $[a, b]$ is said to be absolutely continuous on $[a, b]$ if, given any $\epsilon > 0$, we can find a positive number δ such that for any finite collection of pairwise disjoint intervals $(a_k, b_k) \subset [a, b]$, $k = 1, 2, \ldots, n$, with $\sum (b_k - a_k) < \delta$, we have $\sum |f(b_k) - f(a_k)| < \epsilon$.

The stipulation *finite* may be replaced by *finite or countable*.

Exercise 5.8.2. Verify the following statements.

a. Absolutely continuous functions are uniformly continuous. Hint: Show that f is continuous on the interval $[a, b]$.

b. Absolutely continuous functions are functions of bounded variation and thus differentiable almost everywhere. Hint: The variation of f over an interval of length δ is less than ϵ. Partition $[a, b]$ into subintervals of length less than δ.

c. Continuous functions are not necessarily absolutely continuous. Hint: Recall Billingsley's function (Section 2.8), a continuous nowhere differentiable function.

d. Differentiable functions are not necessarily absolutely continuous, for example,

$$f(x) = \begin{cases} x^2 \sin(\pi/x^2) & 0 < x \le 1, \\ 0 & x = 0. \end{cases}$$

Hint: $x_k = \sqrt{2/(2k+1)}$, $k = 1, 2, \ldots$.

e. Differentiable functions with a bounded derivative are absolutely continuous. Hint:

$$\sum |f(b_k) - f(a_k)| = \sum |f'(c_k)| (b_k - a_k) \le B \sum (b_k - a_k).$$

f. Absolutely continuous functions are differentiable almost everywhere. If $f' = 0$ almost everywhere and f is absolutely continuous, then f is constant.

Theorem 5.8.2. *If f is absolutely continuous on $[a, b]$ (and thus differentiable almost everywhere), and the derivative of f vanishes almost everywhere on $[a, b]$, then f is constant on $[a, b]$.*

Proof. We will show that $f(c) = f(a)$ for any c in the interval $(a, b]$.

Because f is absolutely continuous on $[a, b]$, given an $\epsilon > 0$ there is a $\delta > 0$ so that for any finite or countable collection of pairwise disjoint intervals (a_k, b_k) with length $\sum (b_k - a_k) < \delta$, we have $\sum |f(b_k) - f(a_k)| < \epsilon$.

Let $E = \{x \in (a, c) \mid f'(x) = 0\}$, the interval $[a, c]$ except for a set of measure zero. For each $x \in E$ we have arbitrarily small closed intervals $[x, x + h]$ for which $|f(x + h) - f(x)| < \epsilon h$ because $f'(x) = 0$. This collection of closed intervals is a Vitali cover of E (Section 5.6.1).

We have a finite collection of disjoint closed intervals $[x_1, x_1 + h_1], \ldots,$ $[x_n, x_n + h_n]$, ordered as $a < x_1 < x_1 + h_1 < x_2 < x_2 + h_2 < \cdots < x_n + h_n < c$. That is,

$$(a, c) = (a, x_1) \cup [x_1, x_1 + h_1] \cup (x_1 + h_1, x_2) \cup [x_2, x_2 + h_2]$$
$$\cup \cdots \cup [x_n, x_n + h_n] \cup (x_n + h_n, c)$$

and $\mu(E - \cup[x_k, x_k + h_k]) < \delta$, the δ of absolute continuity. Because

$$(a, c) - \cup[x_k, x_k + h_k] \subseteq ((a, c) - E) \cup (E - \cup[x_k, x_k + h_k])$$

and because $(a, c) - E$ is a set of measure zero, we have

$$\mu((a, c) - \cup[x_k, x_k + h_k])$$
$$= \mu((a_1, x_1) \cup (x_1 + h_1, x_2) \cup \cdots \cup (x_n + h_n, c))$$
$$< \delta.$$

Thus

$$|f(c) - f(a)|$$
$$= |[f(c) - f(x_n + h_n)] + [f(x_n + h_n) - f(x_n)]$$
$$+ \cdots + [f(x_1 + h_1) - f(x_1)] + [f(x_1) - f(a)]|$$
$$\leq (|f(c) - f(x_n + h_n)| + \cdots + |f(x_1) - f(a)|)$$
$$+ (|f(x_1 + h_1) - f(x_1)| + \cdots + |f(x_n + h_n) - f(x_n)|)$$
$$< \epsilon + \epsilon(h_1 + h_2 + \cdots + h_n) < \epsilon(1 + c - a). \quad \square$$

Since the derivative of the Cantor function is zero almost everywhere, and $C(0) = 0$, $C(1) = 1$, the Cantor function — even though continuous and differentiable almost everywhere — is not absolutely continuous.

5.9 What Does Monotonicity Tell Us?

A remarkable theorem (Theorem 5.10.1, whose proof will appear later) tells us that monotone functions are differentiable almost everywhere. To appreciate that monotonicity implies differentiability almost everywhere requires a more detailed analysis of difference quotients

$$\frac{f(y) - f(x)}{y - x}.$$

The *Dini derivates* have proved to be invaluable for such analysis.

5.9.1 Dini Derivates of a Function

Suppose f is defined on an interval containing the point x. We are interested in the four quantities illustrated in Figure 1. They are the limits of the difference quotients at x:

$$D^+ f(x) = \lim_{h \to 0^+} \sup \left\{ \frac{f(y) - f(x)}{y - x}, x < y < x + h \right\},$$

$$D_+ f(x) = \lim_{h \to 0^+} \inf \left\{ \frac{f(y) - f(x)}{y - x}, x < y < x + h \right\},$$

$$D^- f(x) = \lim_{h \to 0^+} \sup \left\{ \frac{f(y) - f(x)}{y - x}, x - h < y < x \right\},$$

$$D_- f(x) = \lim_{h \to 0^+} \inf \left\{ \frac{f(y) - f(x)}{y - x}, x - h < y < x \right\}.$$

The limits always exist (in the extended reals).

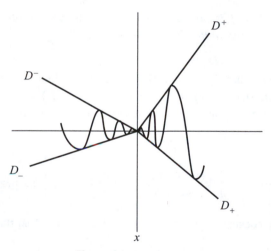

Figure 1. Dini derivates

Exercise 5.9.1. Calculate the Dini derivatives for these functions at $x = 0$.

a. $f(x) = \begin{cases} 1 & x = 0, \\ 0 & x \neq 0, \end{cases}$ $\qquad g(x) = \begin{cases} 0 & x = 0, \\ 1 & x \neq 0, \end{cases}$ $\qquad h(x) = |x|.$

b. $f(x) = \begin{cases} 0 & x = 0, \\ x \sin(\pi/x) & x \neq 0. \end{cases}$

c. Letting $f(0) = 0$,

$$f(x) = \begin{cases} x & x \text{ rational}, \\ -3x & x \text{ irrational}, \end{cases} \qquad \text{for } x > 0,$$

$$f(x) = \begin{cases} -4x & x \text{ rational}, \\ 2x & x \text{ irrational}, \end{cases} \qquad \text{for } x < 0.$$

Exercise 5.9.2.

a. A function is differentiable at x in (a, b) iff all four derivates are finite and equal. Demonstrate.

b. *The Straddle Lemma.* Suppose f is differentiable at a point x in $[a, b]$. Show that for every $\epsilon > 0$ there exists a $\delta_\epsilon(x) > 0$ so that if $u \neq v$ and $x - \delta_\epsilon(x) < u \leq x \leq v < x + \delta_\epsilon(x)$, then

$$\left| f(v) - f(u) - f'(x)(v - u) \right| \leq \epsilon(v - u).$$

Theorem 5.9.1. *If f is monotone increasing on $[a, b]$, then all four derivates are nonnegative and finite almost everywhere.*

Proof. Because $0 \leq D_+ f \leq D^+ f$ and $0 \leq D_- f \leq D^- f$, it is sufficient to show that $D^- f$ and $D^+ f$ are finite almost everywhere. Let $E = \{x \in [a, b] | D^+ f(x) = \infty\}$, and assume $\mu^*(E) = \alpha > 0$. We will arrive at a contradiction after an application of Vitali's Covering Theorem (Section 5.6.1).

For $x \in E$,

$$\lim_{h \to 0^+} \sup \left\{ \frac{f(y) - f(x)}{y - x} : x < y < x + h \right\} = +\infty.$$

Then for any constant K we have a sequence $y_n \to x^+$ so that

$$\frac{f(y_n) - f(x)}{y_n - x} > K.$$

That is, $\{[x, y_n] | x \in E\}$ is a Vitali cover of E.

Thus, we have a finite, disjoint collection, $[x_1, y_1], [x_2, y_2], \ldots, [x_n, y_n]$, so that

$$\mu^* \left(E - \cup_{k=1}^n [x_k, y_k] \right) < \frac{\alpha}{2} \qquad \text{or} \qquad \sum_{k=1}^n (y_k - x_k) > \frac{\alpha}{2}.$$

Then

$$f(b) - f(a) \geq \sum_{k=1}^{n} [f(y_k) - f(x_k)] > K \sum_{k=1}^{n} (y_k - x_k) > K\frac{\alpha}{2}.$$

By choosing K larger than $\left(2(f(b) - f(a))\right)/\alpha$ we have a contradiction.

Complete the argument by showing that $D^- f$ is finite almost everywhere.
□

5.10 Lebesgue's Differentiation Theorem

Finally, as promised, Lebesgue's Differentiation Theorem.

Theorem 5.10.1 (Lebesgue, 1904). *If f is nondecreasing on $[a, b]$, then f is differentiable almost everywhere.*

The proof may be found in Gordon (1994). We include it here as a testament to human ingenuity.

Proof. Since f is nondecreasing, the four Dini derivates are nonnegative, and (by the previous theorem) finite almost everywhere. It will be sufficient to show that the four derivates are equal almost everywhere. Because $0 \leq D_+ f \leq D^+ f$, we will show that the set $E = \{x \in (a, b) \mid D_+ f(x) < D^+ f(x)\}$ has Lebesgue outer measure zero.

In fact, we may reduce the problem further with the observation that

$$E = \cup_{p,q \text{ rational numbers}} \{x \in (a, b) \mid D_+ f(x) < p < q < D^+ f(x)\}.$$

It will be sufficient, then, to show that

$$\mu^*\left(\{x \in (a, b) \mid D_+ f(x) < p < q < D^+ f(x)\}\right) = 0$$

for each pair of rational numbers p, q. Denoting this set by E_{pq}, we assume $\mu^*(E_{pq}) = \alpha > 0$ for some pair of rational numbers p and q. We will arrive at a contradiction.

Given an $\epsilon > 0$, we have an open set O containing E_{pq}. (We may as well assume $O \subset (a, b)$, since (a, b) is an open interval containing E_{pq}.) So $\mu^*(E_{pq}) \leq \mu(O) < \mu^*(E_{pq}) + \epsilon$; that is, $\mu(O) < \alpha + \epsilon$.

For $x \in E_{pq}$, where $D_+ f(x) < p$, we have intervals $[x, y]$ with $y \to x^+$ and $[f(y) - f(x)]/(y - x) < p$. The collection of such intervals forms a Vitali cover of E_{pq}. Thus, we have a finite number of disjoint intervals from this collection, say $[x_1, y_1], [x_2, y_2], \ldots, [x_N, y_N]$, all belonging to O (with $x \in E_{pq} \subset O$, O open), so that $\mu^*(E_{pq} - \cup_{k=1}^N [x_k, y_k]) < \epsilon$, and

$$\sum_{k=1}^N [f(y_k) - f(x_k)] < p \sum_{k=1}^N (y_k - x_k) < p\mu(O) < p(\alpha + \epsilon). \quad (1)$$

Now consider the set $E_{pq} \cap \left(\cup_{k=1}^N [x_k, y_k]\right)$. Because

$$\alpha = \mu^*(E_{pq})$$
$$\leq \mu^* \left(E_{pq} \cap \left(\cup_{k=1}^N [x_k, y_k]\right)\right) + \mu^* \left(E_{pq} - \left(\cup_{k=1}^N [x_k, y_k]\right)\right)$$

we have $\mu^* \left(E_{pq} \cap \left(\cup_{k=1}^N [x_k, y_k]\right)\right) > \alpha - \epsilon$.

A point x in this set belongs to E_{pq} and belongs to *exactly* one of the disjoint intervals $[x_k, y_k]$, with $k = 1, 2, \ldots, N$. We have problems if $x = y_k$, since we want to approach from the right. But there are no such problems if we consider the set $E_{pq} \cap \left(\cup_{k=1}^N (x_k, y_k)\right)$. Furthermore, deletion of the endpoints does not alter the outer measure.

A point in $E_{pq} \cap \left(\cup_{k=1}^N (x_k, y_k)\right)$ belongs to E_{pq} and exactly one of the open intervals (x_k, y_k), say (x_K, y_K). Again we have intervals $[u, v]$ with $v \to u^+$, $[u, v] \subset (x_K, y_K)$, and $[f(v) - f(u)]/(v - u) > q$. The collection of such intervals forms a Vitali cover of $E_{pq} \cap \left(\cup_{k=1}^N (x_k, y_k)\right)$.

Thus we have a finite number of disjoint intervals from this collection, say $[u_1, v_1], [u_2, v_2], \ldots, [u_M, v_M]$, so that

$$\mu^* \left(\left(E_{pq} \cap \left(\cup_{k=1}^N (x_k, y_k)\right)\right) - \cup_{k=1}^M [v_k, u_k]\right) < \epsilon$$

and

$$\sum_{k=1}^M [f(v_k) - f(u_k)] > q \sum_{k=1}^M (v_k - u_k). \quad (2)$$

Since f is nondecreasing, and since each $[u_k, v_k] \subset (x_i, y_i)$ for $k = 1, 2, \ldots, M$ and some $i = 1, 2, \ldots, N$,

$$\sum_{k=1}^M [f(v_k) - f(u_k)] \leq \sum_{k=1}^N [f(y_k) - f(x_k)].$$

From equations 1 and 2 we have

$$q \sum_{k=1}^M (v_k - u_k) < p(\alpha + \epsilon).$$

We observe that

$$\sum_{k=1}^{M}(v_k - u_k) = \sum_{k=1}^{M} \mu([u_k, v_k]) = \mu\left(\cup_{k=1}^{M}[u_k, v_k]\right)$$

$$> \mu^*\left(E_{pq} \cap \left(\cup_{k=1}^{N}(x_k, y_k)\right)\right) - \epsilon > \alpha - 2\epsilon$$

Hence, $q(\alpha - 2\epsilon) < p(\alpha + \epsilon)$, and since ϵ was arbitrary $q\alpha \leq p\alpha$, or $q \leq p$. We have a contradiction to $p < q$. This completes the proof. $\square$

Having an understanding of measurable sets and measurable functions, we are in position to define the Lebesgue integral and discuss Lebesgue integration.

5.11   References

1. Gordon, Russell A. *The Integrals of Lebesgue, Denjoy, Perron, and Henstock.* Providence, R.I.: American Mathematical Society, 1994.

2. Kannan, Rangachary, and Carole King Krueger. *Advanced Analysis on the Real Line.* New York: Springer, 1996.

3. Stromberg, Karl. *An Introduction to Classical Real Analysis.* Belmont, Calif.: Wadsworth, 1981.

CHAPTER **6**

The Lebesgue Integral

As the drill will not penetrate the granite unless kept to the work hour after hour, so the mind will not penetrate the secrets of mathematics unless held long and vigorously to the work. As the sun's rays burn only when concentrated, so the mind achieves mastery in mathematics, and indeed in every branch of knowledge, only when its possessor hurls all his forces upon it. Mathematics, like all the other sciences, opens its doors to those only who knock long and hard. — B. F. Finkel

6.1 Introduction

The culmination of our efforts regarding measure theory, Lebesgue integration, is a mathematical idea with numerous and far-reaching applications. We will confine our remarks to the essential concepts, but the interested reader will be well rewarded by additional efforts.

6.1.1 Lebesgue's Integral

We begin our exploration of Lebesgue's integral (1902) by defining what it means to be *Lebesgue integrable*.

Suppose f is a bounded measurable function on the interval $[a, b]$, so $\alpha < f < \beta$. Partition the range of f: $\alpha = y_0 < y_1 < \cdots < y_n = \beta$, and let $E_k = \{x \in [a, b] \mid y_{k-1} \leq f < y_k\}$, for $k = 1, 2, \ldots, n$.

Form the lower sum $\sum_{k=1}^{n} y_{k-1} \mu(E_k)$ and the upper sum $\sum_{k=1}^{n} y_k \mu(E_k)$ (The terms $\mu(E_k)$ make sense because f is measurable by assumption.) All such sums are between $\alpha(b - a)$ and $\beta(b - a)$.

Now compare the supremum of the lower sums with the infimum of the upper sums over all possible partitions of $[\alpha, \beta]$. If these two numbers are equal, say A, we say f is *Lebesgue integrable* on $[a, b]$, and we write $A = \mathrm{L} \int_a^b f \, d\mu$.

Proposition 6.1.1. *For bounded measurable functions on $[a, b]$, the supremum of the lower sums equals the infimum of the upper sums. The Lebesgue integral always exists.*

We will prove this proposition in four steps.

Step 1. Adding a finite set of points to a partition $\hat{P}$ of $[\alpha, \beta]$ does not decrease the lower sum or increase the upper sum for $\hat{P}$. (So-called "refinements" of $[\alpha, \beta]$ generally increase lower sums and decrease upper sums.)

Step 2. No lower sum can exceed an upper sum. Hint:

$$\sum_{\hat{P}_1} y_{k-1} \mu(E_k) \leq \sum_{\hat{P}_1 \cup \hat{P}_2} z_{i-1} \mu(F_i) \leq \sum_{\hat{P}_1 \cup \hat{P}_2} z_i \mu(F_i) \leq \sum_{\hat{P}_2} s_j \mu(G_j).$$

Step 3. The supremum of the collection of all numbers associated with lower sums is less than or equal to the infimum of the collection of all numbers associated with upper sums.

Step 4. Let $\epsilon > 0$ be given. Construct a partition P^* of $[\alpha, \beta]$ so that $\alpha = y_0^* < y_1^* < \cdots < y_n^* = b$, with $y_k^* - y_{k-1}^* < \epsilon/(b-a)$, for $k = 1, 2, \ldots, n$. Then

$$\alpha(b-a) \leq \sum_{k=1}^{n} y_{k-1}^* \mu(E_k) \leq \sum_{k=1}^{n} y_k^* \mu(E_k) \leq \beta(b-a),$$

and

$$0 \leq \sum_{k=1}^{n} (y_k^* - y_{k-1}^*) \mu(E_k) < \sum_{k=1}^{n} \left(\frac{\epsilon}{b-a}\right) \mu(E_k) = \epsilon.$$

The lower and upper sums for this partition are within ϵ of each other. Thus the supremum of the lower sums and the infimum of the upper sums are within ϵ of each other, and thus because ϵ is arbitrary these numbers are the same.

For a bounded measurable function on an interval $[a, b]$ the Lebesgue integral always exists.

Exercise 6.1.1. Let $L \int_a^b f \, d\mu$ be the Lebesgue integral of the bounded measurable function f. Then for any $\epsilon > 0$ we may construct a partition $\hat{P}$ of $[\alpha, \beta]$ so that the difference of the upper and lower sums for this partition is less than ϵ. Hint: Properties of supremum and infimum.

6.1.2   Young's Approach

Another common approach to the Lebesgue integral is that of William H. Young (1863–1942). Young's method (1905) closely parallels the development of the Riemann integral, where f is assumed only to be bounded and we work with the domain $[a, b]$ instead of the range $[\alpha, \beta]$. Again, we will explore this approach in stages.

First, suppose f is a bounded function on $[a, b]$ with $\alpha < f < \beta$. Partition the domain of f, $[a, b]$, into a finite number of non-overlapping measurable sets E_k, where $\mu(E_i \cap E_j) = 0$ and $E_k \neq \phi$, a so-called measurable partition of $[a, b]$. Pick a point c_k in the measurable set E_k, and form the Lebesgue sums

$$\sum_{k=1}^{n} f(c_k)\mu(E_k), \qquad \text{written } \sum_{P} f(c)\mu(E). \tag{1}$$

These sums satisfy

$$\alpha(b - a) \leq \sum_{P} \inf_{E} f\mu(E) \leq \sum_{P} f(c_k)\mu(E)$$

$$\leq \sum_{P} \sup_{E} f\mu(E) \leq \beta(b - a).$$

Now compare the supremum of the lower Lebesgue sums with the infimum of the upper Lebesgue sums over all measurable partitions of $[a, b]$. If these numbers are the same, say A, we say f is Lebesgue integrable on $[a, b]$, and we write $A = L\int_a^b f \, d\mu$. For E, a measurable subset of $[a, b]$,

$$L\int_E f \, d\mu \equiv L\int_a^b f\chi_E \, d\mu.$$

Before continuing with Young's approach to Lebesgue integrability, some exercises will help clarify a few concepts.

Exercise 6.1.2. Demonstrate the following assertions.

a. Measurable refinements of $[a, b]$ do not decrease lower sums or increase upper sums. Hint: $E_k = F_1 \cup \cdots \cup F_{j_k}$, all sets nonempty measurable subsets of $[a, b]$,

$$\inf_{E_k} f\mu(E_k) \leq \sum \inf_{F_j} f\mu(F_j) \leq \sum \sup_{F_j} f\mu(F_j) \leq \sup_{E_k} f\mu(E_k).$$

b. No lower sum can exceed an upper sum. Hint: $P = \cup_1^n E_k$, $Q = \cup_1^m F_j$; form nm measurable sets $E_k \cap F_j$. Show

$$\sum_P \inf_E f\mu(E) \le \sum_{P \cup Q} \inf_{E \cap F} f\mu(E \cap F) \le \sum_{P \cup Q} \sup_{E \cap F} f\mu(E \cap F)$$

$$\le \sum_Q \sup_F f\mu(F).$$

c. Let $\epsilon > 0$ be given. If we have a partition of $[a, b]$ with the upper and lower sums within ϵ of each other, then the supremum of the upper sums and the infimum of the lower sums are within ϵ of each other. The bounded function f is Lebesgue integrable on $[a, b]$.

d. If the bounded function f is Lebesgue integrable on $[a, b]$, then for any $\epsilon > 0$ we have a partition of $[a, b]$ so that the associated upper and lower sums are within ϵ of each other. Hint: The supremum and infimum properties.

Recall that f's being bounded did not guarantee existence of the Riemann integral — the additional, and sufficient, condition of *continuous almost everywhere* had to be imposed (Theorem 3.6.1). Likewise, in Young's approach an additional and sufficient condition, *measurability of f*, is imposed.

We claim that the bounded function f on $[a, b]$ is Lebesgue integrable on $[a, b]$ iff f is Lebesgue measurable. For the bounded ($\alpha < f < \beta$) measurable function f, we can simply use Section 6.1.1. Hint:

$$\sum y_{k-1} \mu\left(f^{-1}\left([y_{k-1}, y_k)\right)\right) \le \sup_P \sum_P \inf f\mu(E) \le \inf_Q \sum_Q \sup f\mu(F)$$

$$\le \sum y_k \mu\left(f^{-1}\left([y_{k-1}, y_k)\right)\right).$$

For the other direction, suppose the bounded function f is Lebesgue integrable on $[a, b]$. We claim that f is Lebesgue measurable on $[a, b]$. From Exercise 6.1.2, we have partitions P_n of $[a, b]$, with P_n a refinement of P_{n-1}, so that $\sum_{P_n} (\sup f - \inf f)\mu(E) < 1/n$.

Let $\phi_n = \inf f$ and $\psi_n = \sup f$ on P_n. Note that $\phi_{n-1} \le \phi_n \le \psi_n \le \psi_{n-1}$ on $[a, b]$. We have monotone sequences of measurable functions, (ϕ_n), (ψ_n) with $\phi_n \le f \le \psi_n$ on $[a, b]$. Thus $\lim \phi_n = \phi \le f \le \psi = \lim \psi_n$; ϕ and ψ are measurable functions on $[a, b]$. We claim $\phi = \psi$ almost everywhere, in which case $f = \psi$ almost everywhere and measurability of f follows (Section 5.7.2).

Let

$$E = \{x \in [a, b] \mid \psi(x) - \phi(x) > 0\}$$

$$= \cup_m \left\{ x \in [a, b] \mid \psi(x) - \phi(x) > \frac{1}{m} \right\}$$

$$\subseteq \cup_m \left\{ x \in [a, b] \mid \psi_n(x) - \phi_n(x) > \frac{1}{m} \right\}.$$

It will be sufficient to show that $\mu(\{x \in [a, b] \mid \psi_n(x) - \phi_n(x) > 1/m\}$ has measure zero.

By construction, $L \int_a^b (\psi_n - \phi_n) \, d\mu < 1/n$. That is,

$$\frac{1}{m} \mu \left(\left\{ x \in [a, b] \mid \psi_n(x) - \phi_n(x) > \frac{1}{m} \right\} \right) < \frac{1}{n}, \qquad \text{or}$$

$$\mu \left(\left\{ x \in [a, b] \mid \psi_n(x) - \phi_n(x) > \frac{1}{m} \right\} \right) < \frac{m}{n} \qquad \text{for all } n.$$

The argument is complete.

We conclude, for bounded functions on $[a, b]$,

- Riemann integrability iff continuity exists almost everywhere;

- Lebesgue integrability iff measurability holds.

6.1.3 And Another Approach

The reader might prefer a blend of the two approaches just sketched. Assume f is a bounded and *measurable* function in the interval $[a, b]$, and use measurable partitions of $[a, b]$ the domain of f as in Young's approach: $\sum f(c_k)\mu(E_k)$. Complete the details.

6.2 Integrability: Riemann Ensures Lebesgue

We now have an important result to explore. In all that follows, we will assume f is Lebesgue measurable whenever we refer to the Lebesgue integral of f.

Theorem 6.2.1. *All Riemann integrable functions are Lebesgue integrable, and*

$$R \int_a^b f(x) \, dx = L \int_a^b f \, d\mu.$$

We begin by recalling that bounded, continuous almost everywhere functions are measurable. The nonoverlapping intervals $[x_{k-1}, x_k]$ form a measurable partition P of $[a, b]$. We have

$$\sum_P \inf f \Delta x \leq \sup_{\mathcal{P}} \left(\sum_E \inf f \mu(E) \right) \leq \inf_{\mathcal{P}} \left(\sum_E \sup_E f \mu(E) \right)$$

$$\leq \sum_P \sup f \Delta x$$

where $\mathcal{P}$ is the collection of all measurable partitions of $[a, b]$. Complete the argument.

Exercise 6.2.1. Show that Dirichlet's function

$$f(x) = \begin{cases} 1 & x \text{ rational}, \\ 0 & x \text{ irrational}, \end{cases}$$

is Lebesgue integrable on the interval $[0, 1]$.

The problem, posed by Dirichlet and considered by Riemann, of developing an integration process to "integrate" such functions, was solved by Lebesgue approximately 75 years later. We frequently write $L \int_a^b f(x)\, dx$ for $L \int_a^b f\, d\mu$.

For example, consider the function

$$f(x) = \begin{cases} 1/q & x = p/q, \ p, q \text{ relatively prime natural numbers}, \ p < q, \\ 0 & x = 0, 1, \text{ or } x \text{ irrational}. \end{cases}$$

We can show that f is Riemann integrable (Exercise 3.4.2) by Theorem 3.6.1, because f is bounded and continuous on the irrationals, that is, almost everywhere. Thus, f is Lebesgue integrable. Finally, because $f = 0$ almost everywhere,

$$L \int_0^1 f\, d\mu = 0 = R \int_0^1 f(x)\, dx.$$

Some Riemann integrals can be calculated by using the power of the Lebesgue integral.

Exercise 6.2.2.

a. Consider the function (Exercise 3.4.1)

$$f(x) = \begin{cases} 1 & 1/2n < x \leq 1/(2n-1), n = 1, 2, \ldots, \\ 0 & \text{otherwise}. \end{cases}$$

Show that $R \int_0^1 f(x)\, dx = L \int_0^1 f(x)\, dx = \ln 2$.

b. For $0 \le x \le 1$, use Lebesgue's idea of partitioning the range $[0, 1]$ of $\sqrt{x}$ with $E_k = [(k-1)/n, k/n)$, $k = 1, 2, \ldots, n+1$, and show $L \int_0^1 \sqrt{x}\, dx = 2 R \int_0^1 x^2\, dx$ by noting that

$$2 \sum \left(\frac{k-1}{n} \right)^2 \frac{1}{n} \le \sum \frac{(k-1)(2k-1)}{n^3}$$

$$= \sum \frac{k-1}{n} \left[\left(\frac{k}{n} \right)^2 - \left(\frac{k-1}{n} \right)^2 \right]$$

$$< \sum \frac{k}{n} \left[\left(\frac{k}{n} \right)^2 - \left(\frac{k-1}{n} \right)^2 \right] < 2 \sum \left(\frac{k}{n} \right)^2 \frac{1}{n}.$$

6.2.1 Nonnegative Unbounded Measurable Functions

We may extend Lebesgue integrability to nonnegative unbounded measurable functions on $[a, b]$ in several ways. A particularly straightforward approach for nonnegative measurable functions f is to work with "truncations" of f:

$$^k f = \begin{cases} f & 0 \le f \le k, \\ k & f > k. \end{cases}$$

Thus the sequence $\{^k f\}$ of bounded measurable functions converges monotonically to f, and we define the Lebesgue integral of f to be the limit of the monotone sequence $\left\{ L \int_a^b {}^k f\, d\mu \right\}$:

$$L \int_a^b f\, d\mu \equiv \lim_k L \int_a^b {}^k f\, d\mu.$$

Of course the limit may be infinite, but we are not interested in this case. We will say that f is Lebesgue integrable on $[a, b]$ provided that $\lim_k L \int_a^b {}^k f\, d\mu$ is a real number.

Exercise 6.2.3.

a. Show that f is Lebesgue integrable on $[0, 1]$ and $L \int_0^1 f\, d\mu = 2$, given

$$f(x) = \begin{cases} 0 & x = 0, \\ 1/\sqrt{x} & 0 < x \le 1, \end{cases}$$

and so

$$^k f(x) = \begin{cases} 0 & x = 0, \\ 1/\sqrt{x} & 1/k^2 \le x \le 1, \\ k & 0 < x < 1/k^2. \end{cases}$$

b. Show that f is not Lebesgue integrable on $[0, 1]$, given

$$f(x) = \begin{cases} 0 & x = 0, \\ 1/x & 0 < x \le 1. \end{cases}$$

c. Show that f is unbounded on every subinterval of $[0, 1]$, that f is Lebesgue integrable, and that $\text{L} \int_0^1 f \, d\mu = 0$, given

$$f(x) = \begin{cases} q & x = p/q, \ p, q \text{ natural numbers, no common} \\ & \text{factors, } 0 < x < 1, \\ 0 & \text{otherwise, } x = 0, x = 1. \end{cases}$$

6.2.2 Positive and Negative Measurable Functions

What about the case when the measurable function f is both positive and negative? Observe that

$$f = \frac{|f| + f}{2} - \frac{|f| - f}{2} \quad \text{and} \quad |f| = \frac{|f| + f}{2} + \frac{|f| - f}{2}.$$

This allows us to write f as the difference and $|f|$ as the sum of two nonnegative measurable functions. Each of the integrals

$$\text{L} \int_a^b \frac{|f| + f}{2} \, d\mu \quad \text{and} \quad \text{L} \int_a^b \frac{|f| - f}{2} \, d\mu$$

may be calculated by truncation if necessary. If both integrals are finite, we say f is Lebesgue integrable on $[a, b]$.

Note that f is Lebesgue integrable iff $|f|$ is Lebesgue integrable.

Exercise 6.2.4.

a. For

$$f(x) = \begin{cases} -1 & x \text{ rational}, \\ 1 & x \text{ irrational}, \end{cases}$$

show $\text{R} \int_0^1 |f(x)| \, dx = 1$ and $\text{R} \int_0^1 f(x) \, dx$ does not exist. However, both $\text{L} \int_0^1 |f| \, d\mu$ and $\text{L} \int_0^1 f \, d\mu$ exist. The function f is Lebesgue integrable iff $|f|$ is Lebesgue integrable. This is not true for Riemann integrals in general.

b. Show that $\text{L} \int_0^2 f \, d\mu = 0$, given

$$f(x) = \begin{cases} 1/\sqrt{x} & 0 < x \le 1, \\ -1/\sqrt{2 - x} & 1 < x < 2. \end{cases}$$

c. Show that f is differentiable on $[0, 1]$, given

$$f(x) = \begin{cases} x^2 \sin(\pi/x^2) & 0 < x \leq 1, \\ 0 & x = 0. \end{cases}$$

d. For the previous function, show that $L \int_0^1 f' \, d\mu$ does not exist. Hint:

$$L \int_0^1 \left| \frac{1}{x} \cos\left(\frac{\pi}{x^2}\right) \right| \, dx \geq \sum_N \int_{(N-1)\pi}^{N\pi} \left| \frac{\cos(y)}{2y} \right| \, dy.$$

6.2.3 Arbitrary Measurable Subsets

For an arbitrary measurable subset E of $[a, b]$, we may define the Lebesgue integral of f over E, written $L \int_E f \, d\mu$, as $L \int_a^b f \chi_E \, d\mu$, the Lebesgue integral of f times the characteristic function on E, over $[a, b]$.

If f is Lebesgue integrable on $[a, b]$ and g equals f almost everywhere on $[a, b]$, then we claim g is Lebesgue integrable on $[a, b]$ and $L \int_a^b g \, d\mu = L \int_a^b f \, d\mu$.

(We frequently say "sets of measure zero do not affect Lebesgue integrals.") Suppose f and g are nonnegative, and consider their truncations $\{^k f\}$, $\{^k g\}$. The functions $^k f$, $^k g$ are bounded measurable functions, Lebesgue integrable; $^k f = {}^k g$ almost everywhere on $[a, b]$; $^k f \neq {}^k g$ on a set Z of measure zero; and $^k f = {}^k g$ on $[a, b] - Z$.

For the measurable partition $([a, b] - Z) \cup Z$ of $[a, b]$, form the Lebesgue sum (1),

$$(^k g - {}^k f)(c_1)\mu([a, b] - Z) + (^k g - {}^k f)(c_2)\mu(Z)$$
$$= 0 \cdot \mu([a, b] - Z) + (^k g - {}^k f)(c_2) \cdot 0 = 0.$$

We may show $L \int_a^b (^k g - {}^k f) \, d\mu = 0$. Thus

$$L \int_a^b {}^k g \, d\mu = L \int_a^b (^k g - {}^k g) \, d\mu + L \int_a^b {}^k f \, d\mu = L \int_a^b {}^k f \, d\mu,$$

which can be extended with limits.

The corresponding result is not necessarily true for Riemann integrals.

The reader may compare Dirichlet's function with the identically zero function on $[0, 1]$, or Exercise 6.2.3c.

6.2.4   Another Definition of the Lebesgue Integral

We offer an equivalent (and in hindsight more easily generalized) definition of the Lebesgue integral. As before, the explanation has several steps.

Step 1. If ϕ is a nonnegative simple function, $\phi = \sum c_k \chi_{E_k}$, where $E = \cup_1^n E_k$, for E_k mutually disjoint measurable subsets of E and c_k nonnegative real numbers, then

$$\mathrm{L}\int_E f \, d\mu \equiv \sum_1^n c_k \mu(E \cap E_k).$$

Step 2. Recall Theorem 5.7.2, that if f is a nonnegative measurable function defined on a measurable set E, we have a monotonically increasing sequence of simple functions $\{\phi_k\}$ converging to f. Then $\mathrm{L}\int_E f \, d\mu \equiv \lim \mathrm{L}\int_E \phi_k \, d\mu$. (This limit is well defined.) A nonnegative measurable function f is said to be integrable on E if $\mathrm{L}\int_E f \, d\mu$ is finite.

Step 3. For a measurable function f, $f = (|f| + f)/2 - (|f| - f)/2$. If $\mathrm{L}\int_E (|f| + f)/2 \, d\mu$ and $\mathrm{L}\int_E (|f| - f)/2 \, d\mu$ are both finite, then we say f is Lebesgue integrable and define

$$\mathrm{L}\int_E f \, d\mu \equiv \mathrm{L}\int_E \frac{|f| - f}{2} \, d\mu - \mathrm{L}\int_E \frac{|f| + f}{2} \, d\mu.$$

   Note: If $g = f$ almost everywhere, and f is Lebesgue integrable, then g is Lebesgue integrable and $\mathrm{L}\int_E g \, d\mu = \mathrm{L}\int_E f \, d\mu$ (with ϕ simple, $\phi = 0$ almost everywhere, et cetera).

6.3   Convergence Theorems

We now discuss some convergence theorems for the Lebesgue integral, one of their strongest attributes. By definition, we have convergence theorems for special situations, *simple* and *truncations*:

$$\mathrm{L}\int_a^b f \, d\mu \equiv \lim \int_a^b \phi_k \, d\mu \quad \text{or} \quad \mathrm{L}\int_a^b f \, d\mu \equiv \lim \int_a^b {}^k f \, d\mu.$$

We want to replace the special measurable functions ϕ_k or ${}^k f$ with more general measurable functions. This will result in convergence theorems that are more powerful than those for the Riemann integral. Generally speaking, we seek to weaken the uniform convergence requirement that is generally associated with the Riemann integral.

Theorem 6.3.1 (Bounded Convergence). *If $\{f_k\}$ is a uniformly bounded sequence of Lebesgue measurable functions converging pointwise to f almost everywhere on $[a, b]$, then*

$$\lim \mathrm{L}\int_a^b f_k \, d\mu = \mathrm{L}\int_a^b f \, d\mu = \mathrm{L}\int_a^b (\lim f_k) \, d\mu.$$

Proof. Define f to be zero wherever $\lim f_k$ is not f. (Sets of measure zero do not affect the Lebesgue integral.) Because the sequence $\{f_k\}$ is uniformly bounded, say $|f_k| \le B$ on $[a, b]$, and limits of measurable functions are measurable, we have that f is bounded and measurable, and thus Lebesgue integrable on $[a, b]$. So $\mathrm{L}\int_a^b f \, d\mu$ makes sense, as does $\mathrm{L}\int_a^b f_k \, d\mu$.

Let $\epsilon > 0$ be given. By Egoroff's Theorem (5.7.3), we have a subset E of $[a, b]$ so that the sequence $\{f_k\}$ converges uniformly to f on $[a, b] - E$, and $\mu(E) < \epsilon$ where $|f - f_k| \le 2B$. Thus,

$$\left| \mathrm{L}\int_a^b f_k \, d\mu - \mathrm{L}\int_a^b f \, d\mu \right|$$

$$\le \mathrm{L}\int_a^b |f_k - f| \, d\mu$$

$$= \mathrm{L}\int_E |f_k - f| \, d\mu + \mathrm{L}\int_{[a,b]-E} |f_k - f| \, d\mu$$

$$\le 2B\mu(E) + \epsilon\mu([a, b] - E) < 2B\epsilon + \epsilon(b - a)$$

for k sufficiently large, and the argument is complete. $\square$

Exercise 6.3.1.

a. For

$$f_k = \begin{cases} k^2 & 0 < x < 1/k, \\ 0 & \text{otherwise}, \end{cases} \qquad \lim f_k = 0, \qquad \lim \mathrm{L}\int_0^1 f_k = \lim k = \infty.$$

b. Define a sequence $\{f_k\}$ by

$$f_k(x) = \begin{cases} 1 & x = r_1, r_2, \ldots, r_k, \\ 0 & \text{otherwise}, \end{cases}$$

where $r_1, r_2, \ldots$ is an enumeration of the rational numbers in $(0, 1)$. Then Dirichlet's function ($\lim f_k$) is Lebesgue integrable and $\mathrm{L}\int_0^1 f \, d\mu = 0$. The limit of a uniformly bounded monotonically increasing sequence of Riemann integrable functions is not Riemann integrable in general.

c. Revisit Exercise 2.5.2.

d. Revisit Exercise 3.4.1.

6.3.1 Monotone Convergence

Uniform boundedness may be replaced with monotonicity. We will now look at the Lebesgue Monotone Convergence Theorem of Beppo Levi.

Theorem 6.3.2 (Levi, 1906). *If $\{f_k\}$ is a monotone increasing sequence of nonnegative measurable functions converging pointwise to the function f on $[a,b]$, then the Lebesgue integral of f exists and*

$$\mathrm{L}\int_a^b f \, d\mu = \lim \mathrm{L} \int_a^b f_k \, d\mu.$$

Proof. The function f, being the limit of a sequence of measurable functions, is measurable. Because every nonnegative measurable function is the limit of a nondecreasing sequence of simple functions (Theorem 5.7.2), and because its integral is by definition (Section 6.2.4) the limit of the sequence of integrals of the simple functions, we have:

$$0 \leq \phi_{11} \leq \phi_{12} \leq \cdots \leq \phi_{1n} \leq \cdots, \quad \lim_n \phi_{1n} = f_1, \quad \text{and}$$

$$\mathrm{L}\int_a^b f_1 \, d\mu \equiv_D \lim_n \mathrm{L} \int_a^b \phi_{1n} \, d\mu,$$

$$0 \leq \phi_{21} \leq \phi_{22} \leq \cdots \leq \phi_{2n} \leq \cdots, \quad \lim_n \phi_{2n} = f_2, \quad \text{and}$$

$$\mathrm{L}\int_a^b f_2 \, d\mu \equiv_D \lim_n \mathrm{L} \int_a^b \phi_{2n} \, d\mu,$$

$$\vdots$$

$$0 \leq \phi_{m1} \leq \phi_{m2} \leq \cdots \leq \phi_{mn} \leq \cdots, \quad \lim_n \phi_{mn} = f_m, \quad \text{and}$$

$$\mathrm{L}\int_a^b f_m \, d\mu \equiv_D \lim_n \mathrm{L} \int_a^b \phi_{mn} \, d\mu, \dots.$$

Construct a new sequence of simple functions, $\hat{\phi}_k$:

$$\hat{\phi}_1 = \phi_{11},$$
$$\hat{\phi}_2 = \max\{\phi_{12}, \phi_{22}\} \geq \phi_{12} \geq \phi_{11} = \hat{\phi}_1,$$
$$\vdots$$
$$\hat{\phi}_k = \max\{\phi_{1k}, \phi_{2k}, \dots, \phi_{k-1k}, \phi_{kk}\} \geq \hat{\phi}_{k-1}.$$

They are nonnegative and form an increasing sequence converging pointwise to f: $\lim \hat{\phi}_k = \lim f_k = f$. By definition, then, $\mathrm{L}\int_a^b f \, d\mu \equiv_D$

$\lim L \int_a^b \hat{\phi}_k \, d\mu$. However,

$$L \int_a^b \hat{\phi}_k \, d\mu \leq L \int_a^b f_k \, d\mu \leq L \int_a^b f \, d\mu,$$

since $\{f_k\}$ is a monotone increasing sequence converging to f. We may therefore conclude that $\lim L \int_a^b f_k \, d\mu = L \int_a^b f \, d\mu$. $\quad\Box$

Exercise 6.3.2. If $\{f_k\}$ is a monotone sequence of Lebesgue integrable functions on the interval $[a, b]$ converging pointwise to f, $\left| \int_a^b f_1 \, d\mu \right| < \infty$, and $\lim L \int_a^b f_k \, d\mu$ is finite, show that f is Lebesgue integrable on $[a, b]$ and $L \int_a^b f \, d\mu = \lim L \int_a^b f_k \, d\mu$. Hint: $(f_1 - f_k)$.

Exercise 6.3.3. Calculate the following.

a. $L \int_0^1 x^{-1/2} \, dx$. Hint:

$$f_k(x) = \begin{cases} 0 & 0 \leq x < 1/k^2, \\ x^{-1/2} & 1/k^2 < x \leq 1. \end{cases}$$

b. $L \int_0^1 -x \ln(x) \, dx$. Hint:

$$f_k(x) = \begin{cases} -x \ln(x) & 1/k \leq x \leq 1, \\ 0 & 0 < x < 1/k. \end{cases}$$

c. $L \int_0^1 \left(\sum_0^\infty \dfrac{x^{3/2}}{(1+x^2)^n} \right) dx$. Hint: Show

$$\sum_0^\infty \frac{x^{3/2}}{(1+x^2)^n} = \begin{cases} 0 & x = 0, \\ x^{-1/2} + x^{3/2} & 0 < x \leq 1. \end{cases}$$

d. $\displaystyle\sum_0 L \int_0^1 \dfrac{x^\alpha}{(1+x^\beta)^n} \, dx$.

e. $\sum_0 L \int_0^1 x^\alpha (1 - x^\beta)^n \, dx$.

f. $\displaystyle\sum_1 L \int_0^1 \dfrac{x}{[1 + (n-1)x](1+nx)} \, dx$.

6.3.2   Sequential Convergence

Pierre Fatou (1878–1929) gave us a result that will prove useful in the development of Fundamental Theorems of Calculus for the Lebesgue Integral. It is widely known as Fatou's Lemma.

Lemma 6.3.1 (Fatou, 1906). *If $\{f_k\}$ is a sequence of nonnegative Lebesgue integrable functions converging pointwise almost everywhere to f on $[a, b]$, then*

$$\mathrm{L}\int_a^b f \, d\mu \le \liminf \mathrm{L}\int_a^b f_k \, d\mu.$$

Proof. Here is the basic idea: "lim inf" yields monotone increasing sequences that are amenable to the Monotone Convergence Theorem. We have $\underline{f}_1 \equiv \inf\{f_1, f_2, \ldots\}$; $\underline{f}_1$ is measurable, where $0 \le \underline{f}_1 \le f_n$ for all n; and

$$\mathrm{L}\int_a^b \underline{f}_1 \, d\mu \le \inf\left\{\mathrm{L}\int_a^b f_1 \, d\mu, \mathrm{L}\int_a^b f_2 \, d\mu, \ldots\right\}.$$

Continuing, $\underline{f}_k \equiv \inf\{f_k, f_{k+1}, \ldots\}$; and again

$$\mathrm{L}\int_a^b \underline{f}_k \, d\mu \le \inf\left\{\mathrm{L}\int_a^b f_k \, d\mu, \mathrm{L}\int_a^b f_{k+1} \, d\mu, \ldots\right\}.$$

So we have

$$0 \le \underline{f}_1 \le \underline{f}_2 \le \cdots \le \underline{f}_k \le \cdots,$$

$$\mathrm{L}\int_a^b \underline{f}_k \le \inf\left\{\mathrm{L}\int_a^b f_k \, d\mu, \mathrm{L}\int_a^b f_{k+1} \, d\mu, \ldots\right\}.$$

The sequences

$$\left\{\mathrm{L}\int_a^b \underline{f}_k \, d\mu\right\} \quad \text{and} \quad \left\{\inf\left\{\mathrm{L}\int_a^b f_k \, d\mu, \mathrm{L}\int_a^b f_{k+1} \, d\mu, \ldots\right\}\right\}$$

are nonnegative monotone increasing sequences of real numbers (integrability of f_k) and have limits in the extended reals:

$$\lim \mathrm{L}\int_a^b \underline{f}_k \, d\mu \le \liminf\left\{\mathrm{L}\int_a^b f_k \, d\mu, \mathrm{L}\int_a^b f_{k+1} \, d\mu, \ldots\right\}.$$

The Monotone Convergence Theorem 6.3.2 yields

$$\lim \mathrm{L}\int_a^b \underline{f}_k \, d\mu = \mathrm{L}\int_a^b \lim \underline{f}_k \, d\mu = \mathrm{L}\int_a^b f \, d\mu \le \liminf \mathrm{L}\int_a^b f_k \, d\mu. \qquad \square$$

6.3.3   The Dominated Convergence Theorem

Finally, we have a very powerful convergence theorem that replaces mono-tonicity with dominance.

Theorem 6.3.3 (Lebesgue,1910). *Suppose $\{f_k\}$ is a sequence of Lebesgue integrable functions (f_k measurable and $L\int_a^b |f_k|\,d\mu < \infty$ for all k) converging pointwise almost everywhere to the function f on $[a, b]$. Let g be Lebesgue integrable so that $|f_k| \leq g$ for all k on the interval $[a, b]$. Then f is Lebesgue integrable on $[a, b]$ and*

$$L\int_a^b f\,d\mu = \lim L\int_a^b f_k\,d\mu.$$

Proof. Since limits of measurable functions are measurable by Theorem 5.7.1, and since sets of measure zero do not affect Lebesgue integrability, we may assume f is real-valued and measurable, and that $\lim f_k = f$ on $[a, b]$.

Construct two monotone sequences $\{\underline{f}_k\}$ and $\{\overline{f}_k\}$, with

$$\underline{f}_k = \inf\{f_k, f_{k+1}, \dots\} \quad \text{and} \quad \overline{f}_k = \sup\{f_k, f_{k+1}, \dots\}.$$

Because $-g \leq \underline{f}_k \leq f_k \leq \overline{f}_k \leq g$, the functions $\underline{f}_k$, $\overline{f}_k$, and f are integrable. Furthermore, since

$$0 \leq g + \underline{f}_k \leq g + \underline{f}_{k+1} \leq 2g \quad \text{and} \quad 0 \leq g - \overline{f}_k \leq g - \overline{f}_{k+1} \leq 2g,$$

the sequences $\{g + \underline{f}_k\}$ and $\{g - \overline{f}_k\}$ are monotone increasing sequences of integrable functions with $\lim(g + \underline{f}_k) = g + f$ and $\lim(g - \overline{f}_k) = g - f$.

Application of the Monotone Convergence Theorem yields

$$L\int_a^b (g + f)\,d\mu = L\int_a^b g\,d\mu + L\int_a^b f\,d\mu = \lim L\int_a^b (g + \underline{f}_k)\,d\mu$$

$$= L\int_a^b g\,d\mu + \lim \int_a^b \underline{f}_k\,d\mu.$$

That is, $L\int_a^b f\,d\mu = \lim \int_a^b \underline{f}_k\,d\mu$. Similarly, $L\int_a^b f\,d\mu = \lim \int_a^b \overline{f}_k\,d\mu$.

With the observation that $\underline{f}_k \leq f_k \leq \overline{f}_k$, and all functions being inte-grable, we have

$$L\int_a^b \underline{f}_k\,d\mu \leq L\int_a^b f_k\,d\mu \leq L\int_a^b \overline{f}_k\,d\mu \qquad \text{for all } k.$$

Thus $L\int_a^b f\,d\mu = \lim \int_a^b f_k\,d\mu$, and the proof is complete. $\square$

Exercise 6.3.4.

a. Show $\lim_{k} L \int_0^1 \dfrac{kx}{1+k^2x^2}\, dx = 0$. Hint. There are many approaches.
Let $g = \frac{1}{2}$ on $[0, 1]$.

b. Evaluate $\lim_{k} L \int_0^1 \dfrac{k^{3/2}x}{1+k^2x^2}\, dx = 0$. Hint.

$$g(x) = \begin{cases} 0 & x = 0, \\ x^{-1/2} & 0 < x \le 1. \end{cases}$$

c. Show $\lim_{k} L \int_{-2}^2 \dfrac{x^{2k}}{1+x^{2k}}\, dx = 2$.

d. Evaluate $L \int_0^1 \left(\sum \dfrac{x^n}{n} \right) dx$.

e. Show $L \int_0^1 \dfrac{1}{1-x} \ln\left(\dfrac{1}{x} \right) dx = \dfrac{\pi^2}{6}$.

f. Show $L \int_0^1 \dfrac{x^{\alpha-1}}{1+x^\beta}\, dx = \dfrac{1}{\alpha} - \dfrac{1}{\alpha+\beta} + \dfrac{1}{\alpha+2\beta} - \dfrac{1}{\alpha+3\beta} + \cdots$,
where $\alpha, \beta > 0$.

6.3.4 Interchanging $\sum$ and $\int$

Suppose $\{f_k\}$ is a sequence of Lebesgue integrable functions defined on $[a, b]$. If the series $\sum \int_a^b |f_k|\, d\mu$ converges, we can show that the series $\sum f_k$ converges almost everywhere on $[a, b]$ to a Lebesgue integrable function and

$$\sum L \int_a^b f_k\, d\mu = L \int_a^b \sum f_k\, d\mu.$$

By the Monotone Convergence Theorem 6.3.2,

$$L \int_a^b \sum |f_k|\, d\mu = \sum L \int_a^b |f_k|\, d\mu < \infty.$$

Because $\sum |f_k|$ is Lebesgue integrable on $[a, b]$, it is thus finite almost everywhere on $[a, b]$. We have that $\sum f_k$ converges almost everywhere on $[a, b]$ and $|\sum f_k| \le \sum |f_k|$.

We claim that $\sum f_k$ is integrable on $[a, b]$. Let $g_n = \sum_1^n f_k$. Then $|g_n| \leq \sum |f_k|$. We apply the Dominated Convergence Theorem 6.3.3:

$$\mathrm{L} \int_a^b \left(\sum f_k \right) d\mu = \mathrm{L} \int_a^b (\lim g_n) \, d\mu = \lim_n \mathrm{L} \int_a^b g_n \, d\mu$$

$$= \lim_n \sum_1^n \int_a^b f_k \, d\mu = \sum_1^\infty \int_a^b f_k \, d\mu.$$

6.4   Fundamental Theorems for the Lebesgue Integral

What remains to explore? We have reached the Fundamental Theorems of Calculus for the Lebesgue integral. Now we can resolve the crucial questions:

- Does $\mathrm{L} \int_a^b f' \, d\mu = f(b) - f(a)$?

- When is $\left(\mathrm{L} \int_a^x f \, d\mu \right)' = f$?

6.4.1   Properties of the Indefinite Integral

We will begin by examining properties of the indefinite integral. That is, for a Lebesgue integrable function f on the interval $[a, b]$, we will define a new function F on $[a, b]$ by

$$F(x) = \mathrm{L} \int_a^x f \, d\mu = \mathrm{L} \int_a^x f(t) \, dt.$$

What are the properties of this function? Is F continuous? Differentiable?...

For the following discussion, assume f is Lebesgue integrable on $[a, b]$; that is, f is Lebesgue measurable and $\mathrm{L} \int_a^b |f| \, d\mu < \infty$. Let $F(x) = \mathrm{L} \int_a^x f(t) \, dt$. We will investigate seven properties of the indefinite integral.

Property 1. *If f is bounded on $[a, b]$, then F is continuous on $[a, b]$.* (In fact, F is Lipschitz.)

Hint: With $a \leq x < y \leq b$,

$$|F(y) - F(x)| = \left| \mathrm{L} \int_x^y f(t) \, dt \right| \leq B(y - x) \qquad \text{for } |f| \leq B \text{ on } [a, b].$$

Property 2. *F is continuous on $[a, b]$ whether f is bounded or not.*

Hint: Truncation. Suppose f is unbounded and nonnegative. Then

$$0 \leq \lim L \int_a^b {}^k f \, d\mu = L \int_a^b f \, d\mu.$$

Let $\epsilon > 0$ be given. We have a natural number K so that

$$L \int_a^b f \, d\mu - \epsilon < \int_a^b {}^k f \, d\mu \leq L \int_a^b f \, d\mu, \qquad \text{when } k \geq K.$$

Now, $F^K(x) = L \int_a^x {}^K f(t) \, dt$ is continuous by Property 1. We have a δ so that for $y \in (x - \delta, x + \delta) \cap [a, b]$, the difference $\left| F^K(y) - F^K(x) \right| < \epsilon$. But

$$|F(y) - F(x)| \leq \left| F(y) - F^K(y) \right| + \left| F^K(y) - F^K(x) \right| + \left| F^K(x) - F(x) \right|$$

$$\leq L \int_a^b (f - {}^K f) \, d\mu + \epsilon + L \int_a^b (f - {}^K f) \, d\mu < 3\epsilon,$$

where $y \in (x - \delta, x + \delta) \cap [a, b]$. The reader may remove the requirements that f be nonnegative with the usual $f = (|f| + f)/2 - (|f| - f)/2$.

Because F is continuous on $[a, b]$, we may conclude that F is uniformly continuous on $[a, b]$.

Property 3. *Given an $\epsilon > 0$, we have a δ so that if E is any measurable subset of $[a, b]$ with $\mu(E) < \delta$, $L \int_E |f| \, d\mu < \epsilon$.*

We have replaced the small interval of Property 2 with a small measurable subset of $[a, b]$. Start by assuming f is bounded on $[a, b]$ so that $|f| \leq B$. Then $L \int_E |f| \, d\mu \leq B\mu(E)$. Let $\delta < \epsilon/B$. For f unbounded and nonnegative,

$$L \int_E f \, d\mu = L \int_E (f - {}^K f) \, d\mu + L \int_E {}^K f \, d\mu$$

$$\leq L \int_a^b (f - {}^K f) \, d\mu + L \int_E {}^K f \, d\mu.$$

Complete the argument using the Lebesgue Monotone Convergence Theorem.

Property 4. *If $F(x) = \int_a^x f(t) \, dt = 0$ for all x in $[a, b]$, then $f(t) = 0$ almost everywhere in $[a, b]$.*

Suppose the set $\{x \in [a, b] \mid f(x) > 0\} = \cup \{x \in [a, b] \mid f(x) > 1/n\}$ has positive measure. Then for some N, we have a closed subset F of

$\{x \in [a,b] \mid f(x) > 1/N\}$ (by Theorem 5.6.1), with $\mu(F) > 0$. Then $[a,b] - F = \cup(a_k, b_k)$, so

$$F(b) = L\int_a^b f \, d\mu = L\int_F f \, d\mu + L\int_{[a,b]-F} f \, d\mu$$

$$= L\int_F f \, d\mu + \sum\left[L\int_a^{b_k} f \, d\mu - L\int_a^{a_k} f \, d\mu\right]$$

$$= L\int_F f \, d\mu > \frac{1}{N}\mu(F) > 0.$$

But $F(b) = 0$ by assumption. Complete the argument.

Property 5. *Assuming f is nonnegative, F is nondecreasing on $[a,b]$.*

In fact, since $f = (|f| + f)/2 - (|f| - f)/2$, F is the difference of two monotone functions. That is, F is a function of bounded variation, and thus F is differentiable almost everywhere on $[a,b]$ by Lebesgue's Differentiation Theorem 5.10.1. So, what is F'?

Property 6. *If f is continuous at x_0, a point in $[a,b]$, then F is differentiable at x_0 and $F'(x_0) = f(x_0)$.*

Review the Fundamental Theorem 3.7.2 and supply the details of the argument. Hint:

$$\left|\frac{F(x_0 + h) - F(x_0)}{h} - f(x_0)\right| = \left|\frac{1}{h}L\int_{x_0}^{x_0+h}(f - f(x_0)) \, d\mu\right|.$$

Property 7. *If f is bounded on $[a,b]$, with $|f| \leq B$, then $F'(x) = f(x)$ almost everywhere.*

We will show first that $L\int_a^x (F' - f) \, d\mu = 0$ for all x in $[a,b]$.

Because F is differentiable almost everywhere, $\lim n[F(x + 1/n) - F(x)] = F'(x)$ for x a point of differentiability of F, almost everywhere. Then $F(x + 1/n)$ makes sense on $[a, x]$ if we extend F to $[a, x + 1]$ by $F(t) = F(x)$, for $x \leq t \leq t + 1$. Since sets of measure zero do not affect the Lebesgue integral, we have

$$L\int_a^x (F' - f) \, d\mu = L\int_a^x \left(\lim n\left[F\left(t + \frac{1}{n}\right) - F(t)\right] - f\right) dt.$$

Can we move "lim" outside the integral? What do our convergence theorems demand? We could use the Bounded Convergence Theorem 6.3.1 — provided we have a uniform bound for the sequence $(n[F(x + 1/n) - F(x)])$.

We do. Since $|f| \leq B$,

$$\left| n \left[F \left(x + \frac{1}{n} \right) - F(x) \right] \right| \leq n \cdot \mathrm{L} \int_x^{x+1/n} |f| \, d\mu \leq B.$$

So, invoking Cauchy integrals, we calculate

$$
\begin{aligned}
\mathrm{L} \int_a^x \lim n &\left[F \left(t + \frac{1}{n} \right) - F(t) \right] dt \\
&= \lim \mathrm{L} \int_a^x n \left[F \left(t + \frac{1}{n} \right) - F(t) \right] dt \\
&= \lim \left[n \cdot \mathrm{C} \int_a^x F \left(t + \frac{1}{n} \right) dt - n \cdot \mathrm{C} \int_a^x F(t) \, dt \right] \\
&= \lim \left[n \cdot \mathrm{C} \int_x^{x+1/n} F(t) \, dt - n \cdot \mathrm{C} \int_a^{a+1/n} F(t) \, dt \right] \\
&= \lim \left[n \cdot F(x) \cdot \frac{1}{n} - n \cdot F(\xi_n) \cdot \frac{1}{n} \right], \qquad a < \xi_n < a + \frac{1}{n}, \\
&= F(x) - F(a).
\end{aligned}
$$

Now use Property 4.

Exercise 6.4.1. Let

$$f(x) = \begin{cases} x & x \text{ irrational,} \\ 0 & x \text{ rational.} \end{cases}$$

Calculate $F(x) = \mathrm{L} \int_0^x f(t) \, dt$ and its derivative.

We have shown that, for f measurable and bounded, $\left(\mathrm{L} \int_a^x f \, d\mu \right)' = f$ almost everywhere. We can, in fact, dispose with the requirement that f be bounded.

6.4.2   A Fundamental Theorem for the Lebesgue Integral

Theorem 6.4.1 (Lebesgue, 1904). *Given that f is Lebesgue integrable on $[a, b]$, define a function f on $[a, b]$ by $F(x) = \mathrm{L} \int_a^x f \, d\mu$. Then F is absolutely continuous on $[a, b]$ and $F' = f$ almost everywhere on $[a, b]$.*

Proof. The absolute continuity of F on the interval $[a, b]$ will follow from Property 3 of indefinite integrals (Section 6.4.1). Given an $\epsilon > 0$, we have a $\delta > 0$ so that if E is any measurable subset of $[a, b]$ with $\mu(E) < \delta$, then $\mathrm{L} \int_E |f| \, d\mu < \epsilon$.

Let (a_k, b_k) be a finite collection of pairwise disjoint intervals with length, $\sum(b_k - a_k)$, less than δ. Then

$$\sum |F(b_k) - F(a_k)| \leq L \int_{\cup[a_k, b_k]} |f| \, d\mu < \epsilon.$$

So F is absolutely continuous on $[a, b]$ and thus differentiable almost everywhere.

We have shown the validity of the second conclusion ($F' = f$ almost everywhere on $[a, b]$), with the assumption that f is bounded (Section 6.4.1).

Since $f = [(|f| + f)/2] - [(|f| - f)/2]$, the difference of two nonnegative measurable functions, we assume f is nonnegative, and we begin with the sequence of bounded, measurable functions, the truncations $\{^k f\}$ of f.

The function $L \int_a^x (f - {}^k f) \, d\mu$ is nonnegative and nondecreasing on $[a, b]$. By Lebesgue's Differentiation Theorem 5.10.1, this function has a nonnegative derivative almost everywhere, and by Property 7, $\left(L \int_a^x {}^k f \, d\mu\right)' = {}^k f$ almost everywhere. So,

$$0 \leq \left(L \int_a^x (f - {}^k f) \, d\mu\right)' = \left(L \int_a^x f \, d\mu\right)' - \left(L \int_a^x {}^k f \, d\mu\right)'$$

$$= \left(L \int_a^x f \, d\mu\right)' - {}^k f$$

almost everywhere.

Thus $\left(L \int_a^x f \, d\mu\right)' - f = F' - f \geq 0$ almost everywhere, and $L \int_a^x (F' - f) \, d\mu \geq 0$.

On the other hand,

$$L \int_a^x (F' - f) \, d\mu = L \int_a^x F' \, d\mu - L \int_a^x f \, d\mu.$$

Now, F is a nondecreasing continuous function (by Properties 2 and 5). So F is Cauchy integrable. Furthermore,

$$0 \leq C \int_a^x n \left[F\left(t + \frac{1}{n}\right) - F(t) \right] dt$$

$$= n \cdot \left[C \int_x^{x+1/n} F(t) \, dt - C \int_a^{a+1/n} F(t) dt \right]$$

$$\leq F(x) - F(a) = \int_a^x f \, d\mu,$$

since $F(t) = F(x)$, $x \leq t \leq x + 1$ and $F(a) \leq F(t)$, for $a \leq t$.

The sequence $\{n[F(x + 1/n) - F(x)]\}$ is a sequence of nonnegative Lebesgue integrable functions whose limit is $F'(x)$ almost everywhere. Applying Fatou's Lemma 6.3.1, we have

$$\text{L} \int_a^x F' \, d\mu = \text{C} \int_a^x \lim \left[n \left(F \left(t + \frac{1}{n} \right) - F(t) \right) \right] dt$$

$$\leq \liminf \text{C} \int_a^x n \left[F \left(t + \frac{1}{n} \right) - F(t) \right] dt$$

$$\leq F(x) - F(a) = \text{L} \int_a^x f \, d\mu.$$

That is, $\text{L} \int_a^x (F' - f) \, d\mu \leq 0$ for all x in $[a, b]$, and thus $F' - f = 0$ almost everywhere, by Property 4.

Demonstrate the case when f is nonpositive.

So, $F' = f$ almost everywhere on $[a, b]$. This completes the proof. $\square$

Compare the Fundamental Theorems 2.4.1, 3.7.2, and 8.8.1.

As for $\text{L} \int_a^b F' \, d\mu$, it would be wonderful if $\text{L} \int_a^b F' \, d\mu = F(b) - F(a)$, but this is not true. Let C denote the Cantor function. Then $C' = 0$ almost everywhere on $[0, 1]$. Thus,

$$0 = \text{L} \int_0^1 C' \, d\mu < 1 = C(1) - C(0).$$

6.4.3   The Other Fundamental Theorem

Are there additional conditions we might impose on the derivative that would yield $\text{L} \int_a^b F' \, d\mu = F(b) - F(a)$? This question was the impetus for Lebesgue's development of his integration process.

Theorem 6.4.2. *If F is a differentiable function, and the derivative F' is bounded on the interval $[a, b]$, then F' is Lebesgue integrable on $[a, b]$ and*

$$\text{L} \int_a^x F' \, d\mu = F(x) - F(a)$$

for x in the interval $[a, b]$.

Proof. Because

$$F'(x) = \lim_{h_n \to 0} \left[\frac{F(x + h_n) - F(x)}{h_n} \right]$$

for any sequence of real numbers $\{h_n\}$ converging to zero, and because limits of measurable functions are measurable, we may conclude the derivative is measurable, and it is bounded by assumption. Thus F' is Lebesgue integrable on $[a, b]$.

Because $\lim n[F(x + 1/n) - F(x)] = F'(x)$, it is natural to consider the sequence of functions $\{n[F(x + 1/n) - F(x)]\}$, with F extended to the interval $[a, b + 1]$ by $F(x) = F(b) + F'(b)(x - b)$, where $b \le x \le b + 1$. This sequence of measurable functions is uniformly bounded:

$$\left| n\left[F\left(t + \frac{1}{n} \right) - F(t) \right] \right| = |F'(c)| \le B, \qquad \text{for } t < c < t + \frac{1}{n},$$

by the mean value theorem for derivatives and the assumption that the derivative is bounded.

We may use the Bounded Convergence Theorem 6.3.1 for our calculation:

$$
\begin{aligned}
\text{L} \int_a^x F' \, d\mu &= \text{L} \int_a^x \lim n \left[F\left(t + \frac{1}{n} \right) - F(t) \right] dt \\
&= \lim \text{R} \int_a^x n \left[F\left(t + \frac{1}{n} \right) - F(t) \right] dt \\
&= \lim n \, \text{R} \int_a^x F\left(t + \frac{1}{n} \right) dt - \lim n \, \text{R} \int_a^x F(t) \, dt \\
&= \lim n \, \text{R} \int_x^{x+1/n} F(t) \, dt - \lim n \, \text{R} \int_a^{a+1/n} F(t) \, dt \\
&= F(x) - F(a).
\end{aligned}
$$

The proof is complete. $\square$

So what remains? We would like to remove the requirement that the derivative is bounded. This cannot be done with the Lebesgue integration process. Recall Exercise 6.2.4. (A solution will appear in Chapter 8.)

6.4.4 The Bounded Variation Condition

We showed that if F has a derivative and the derivative is bounded, then $\text{L} \int_a^b F' \, d\mu = F(b) - F(a)$. The assumption that F is differentiable with a bounded derivative implies that F is absolutely continuous on $[a, b]$; recall Exercise 5.8.2.

Would bounded variation of F be sufficient to have $L \int_a^b F' \, d\mu = F(b) - F(a)$? No; for example

$$F(x) = \begin{cases} 1 & 0 \le x \le 1, \\ 2 & 1 < x \le 2; \end{cases}$$

$$F' = 0 \text{ almost everywhere;}$$

$$L \int_0^1 F' \, d\mu = 0 < 1 = F(2) - F(0).$$

The requirement we are looking for is that F is absolutely continuous on $[a, b]$.

6.4.5   Another Fundamental Theorem of Calculus

Let's look at one more Fundamental Theorem of Calculus for the Lebesgue integral.

Theorem 6.4.3 (Lebesgue, 1904). *If F is absolutely continuous on $[a, b]$, then F' is Lebesgue integrable, and*

$$L \int_a^x F' \, d\mu = F(x) - F(a) \qquad \text{for } x \text{ in } [a, b].$$

Proof. We know that F is of bounded variation on $[a, b]$. Thus, $F = F_1 - F_2$ with F_1, F_2 monotone increasing functions and F' exists almost everywhere. Since $|F'| \le F_1' + F_2'$ almost everywhere,

$$L \int_a^b |F'| \, d\mu \le L \int_a^b F_1' \, d\mu + L \int_a^b F_2' \, d\mu$$
$$\le F_1(b) - F_1(a) + F_2(b) - F_2(a)$$

by the proof of Theorem 6.4.1. Thus F' is integrable.

Let $G(x) = L \int_a^x F' \, d\mu$. By Theorem 6.4.1, G is absolutely continuous, so $F - G$ is absolutely continuous and $(F - G)' = F' - G' = 0$ almost everywhere. By Theorem 5.8.2, $F - G$ is constant. Because $G(a) = 0$, we have

$$F(x) - F(a) = L \int_a^x F' \, d\mu. \qquad \square$$

Compare Theorems 2.3.1, 3.7.1, and 8.7.3.

6.4.6 Comments

For the absolute continuity of ϕ, refer to Definition 5.8.2.

Theorem 6.4.4. *If f is continuous and ϕ is absolutely continuous on an interval $[a, b]$, then the Riemann–Stieljes integral of f with respect to ϕ is just the Lebesgue integral of $f\phi'$ on the interval $[a, b]$:*

$$\text{R-S} \int_a^b f(x)\, d\phi(x) = \text{L} \int_a^b f\phi'\, d\mu.$$

Proof. How can we show this? First, since absolutely continuous functions are of bounded variation, the existence of R-S $\int_a^b f(x)\, d\phi(x)$ may be shown by Theorem 4.4.1. In fact,

$$\left(\text{R-S} \int_a^x f(t)\, d\phi(t) \right)' = f\phi'$$

almost everywhere.

Second, $f\phi'$ is Lebesgue integrable on $[a, b]$ and $\text{L} \int_c^d \phi'\, d\mu = \phi(d) - \phi(c)$, for $[c, d] \subset [a, b]$, by Theorem 6.4.3.

Next we have

$$\left| \text{R-S} \int_a^b f(x)\, d\phi(x) - \text{L} \int_a^b f\phi'\, d\mu \right|$$

$$\leq \left| \text{R-S} \int_a^b f(x)\, d\phi(x) - \sum f(c_k)\Delta\phi \right|$$

$$+ \left| \sum f(c_k)\Delta\phi - \text{L} \int_a^b f\phi'\, d\mu \right|.$$

Also,

$$\left| \sum f(c_k)\left[\phi(x_k) - \phi(x_{k-1})\right] - \text{L} \int_a^b f\phi'\, d\mu \right|$$

$$= \left| \sum \left[\text{L} \int_{x_{k-1}}^{x_k} (f(c_k) - f(t))\, \phi'(t)\, dt \right] \right|$$

$$\leq \sum \text{L} \int_{x_{k-1}}^{x_k} |f(c_k) - f(t)|\, |\phi'(t)|\, dt.$$

We may use uniform continuity of f and integrability of ϕ' (Theorem 6.4.3) to finish the argument. □

Theorem 6.4.4 is not true if we assume only that ϕ is of bounded variation. For example, let $f = \phi = C$, the Cantor function. Then by Theorem 4.4.1 and Exercise 4.4.1,

$$\text{R-S}\int_0^1 C \, dC = \frac{1}{2}, \quad \text{but} \quad \text{L}\int_0^1 CC' \, d\mu = 0.$$

6.5 Spaces

We will now discuss one of the most exciting and profound applications of the Lebesgue integral: L-p spaces, written L^p.

We have dealt with various collections of functions — Cauchy integrable functions, absolutely continuous functions, Riemann integrable functions, and many more. In many cases we can define a distance between functions that satisfies all the usual requirements imposed when dealing with real numbers, and which satisfies our intuitive requirements for how distance operates. That is:

1. The distance from any object to itself should be zero.

2. The distance between different objects should be positive.

3. The distance between object 1 and object 2 should be the same as the distance between object 2 and object 1.

4. Finally, just as real numbers satisfy a triangle inequality (that is, $|a - c| \leq |a - b| + |b - c|$), the distance between any two objects should not exceed the sum of the distances between each of these objects and a third object.

6.5.1 Metric Space

More formally, we have the idea of a *metric space*, defined as follows.

Let X be a nonempty set. A metric ρ on X is a real-valued function with domain $X \times X$ that satisfies the following criteria:

1. $\rho(x, x) = 0$, $x \in X$.

2. $\rho(x, y) > 0$, $x, y \in X$ and $x \neq y$.

3. $\rho(x, y) = \rho(y, x)$, $x, y \in X$.

4. $\rho(x, y) \leq \rho(x, z) + \rho(z, y)$ (triangle inequality).

The set X together with the metric ρ (distance function) is called a *metric space*, denoted by (X, ρ). Let's look at some examples of metric spaces.

Example 6.5.1. For real numbers x, y, $\rho(x, y) = |x - y| : (R, |x - y|)$.

Example 6.5.2. For continuous functions on $[0, 1]$, $C[0, 1]$,

$$\rho(x, y) = \max_{0 \le t \le 1} |x(t) - y(t)| : \left(C[0, 1], \max_{0 \le t \le 1} |x - y| \right).$$

In this second instance, the reader may show we have a metric space. For example, invoking the triangle inequality, we have

$$|x(t) - y(t)| \le |x(t) - z(t)| + |z(t) - y(t)|, \qquad \text{for } 0 \le t \le 1,$$

because $x(t)$, $y(t)$, and $z(t)$ are real numbers. Certainly

$$|x(t) - y(t)| \le \max_{0 \le t \le 1} |z(t) - x(t)| + \max_{0 \le t \le 1} |z(t) - y(t)|,$$

and the result follows.

Or does it? Recall that x, y are continuous functions, $x - y$ is a continuous function, $|x - y|$ is a continuous function, and continuous functions assume a maximum on compact sets. Tighten up the argument.

Example 6.5.3. Consider $C[0, 1]$, the same collection as in the preceding example, but with a different metric: $\rho(x, y) = C \int_0^1 |x(t) - y(t)| \, dt$.

Because we are dealing with the Cauchy integral, it is important that $|x(t) - y(t)|$ be a continuous function on $[0, 1]$. It is.

Now, if $x \ne y$, is $C \int_0^1 |x(t) - y(t)| \, dt > 0$? Let's see; $|x(t_0) - y(t_0)| > 0$, so we have a neighborhood about $t_0 \ldots$. And the triangle inequality entails

$$C \int_0^1 |x(t) - y(t)| \, dt \le C \int_0^1 |x(t) - z(t)| \, dt + C \int_0^1 |z(t) - y(t)| \, dt.$$

But $|x(t) - y(t)| \le |x(t) - z(t)| + |z(t) - y(t)|$ holds for all t in $[0, 1]$. We can use the monotonicity property of integrals.

In this example it was important that, given objects in our collection, the difference of such objects belongs to the collection. Again, we write

$$\left(C[0, 1], \int_0^1 |x(t) - y(t)| \, dt \right).$$

Example 6.5.4. Consider $R[0, 1]$, the collection of Riemann integrable functions on $[0, 1]$, with $\rho(x, y) \equiv R \int_0^1 |x(t) - y(t)| \, dt$.

Here, we have that x and y are bounded and continuous almost everywhere, $x - y$ is bounded and continuous almost everywhere, $|x - y|$ is bounded and continuous almost everywhere. So $R \int_0^1 |x(t) - y(t)| \, dt$ makes sense.

If $x \neq y$, is $R \int_0^1 |x(t) - y(t)| \, dt > 0$? Suppose $x \equiv 0$ in $[0, 1]$, with $y = 1$, $t = \frac{1}{2}$, 0 otherwise. Then $x \neq y$, but $R \int_0^1 |x(t) - y(t)| \, dt = 0$. We have a problem, unless we agree to identify functions that are equal almost everywhere. This we will do.

The "points" in our space $R[0, 1]$ are actually classes of functions, the functions in a particular class differing from each other on a set of measure zero. However, we will follow tradition and talk about a *function x from the metric space*, in this case $R[0, 1]$, when technically speaking, we should talk about a *representation*.

We have $\left(R[0, 1], R \int_0^1 |x(t) - y(t)| \, dt \right)$.

Example 6.5.5. Consider $L^1[0, 1]$, the collection of Lebesgue measurable functions x on $[0, 1]$ with $L \int_0^1 |x| \, d\mu < \infty$.

The reader can show we have a metric.
The metric space $\left(L^1[0, 1], L \int_0^1 |x - y| \, d\mu \right)$ is also called L^1.

Example 6.5.6. Consider $L^2[0, 1]$, the collection of Lebesgue measurable functions x on $[0, 1]$ with $L \int_0^1 |x|^2 \, d\mu < \infty$.

In this instance, we have a metric space with

$$\rho(x, y) = \sqrt{L \int_0^1 |x - y|^2 \, d\mu}.$$

Is

$$\sqrt{L \int_0^1 |x - y|^2 \, d\mu} \leq \sqrt{L \int_0^1 |x - z|^2 \, d\mu} + \sqrt{L \int_0^1 |z - y|^2 \, d\mu}$$

true? This inequality would follow from $|x - y|^2 \leq |x - z|^2 + |z - y|^2$.

But first, is $x - y$ a member of $L^2[0, 1]$? That is, given $x, y \in L^2[0, 1]$ does it follow that $L \int_0^1 |x + y|^2 \, d\mu < \infty$? We calculate

$$|x + y|^2 \leq (|x| + |y|)^2 \leq (2 \max\{|x|, |y|\})^2$$
$$\leq 4 \max\{|x|^2, |y|^2\} \leq 4(|x|^2 + |y|^2).$$

So, $|x + y|^2$ is measurable and $L \int_0^1 |x + y|^2 \, d\mu < \infty$. Thus it makes sense to talk about $x - y$, $x - z$, $z - y$ in this space of functions $L^2[0, 1]$. Still, the triangle inequality remains.

However,

$$|x - y|^2 = |x - z + z - y|^2 \leq |x - z + z - y| (|x - z| + |z - y|)$$
$$= |x - y||x - z| + |x - y||y - z|.$$

Thus, we have

$$\sqrt{L \int_0^1 |x - y|^2 \, d\mu} \leq \sqrt{L \int_0^1 |x - y||x - z| \, d\mu}$$
$$+ \sqrt{L \int_0^1 |x - y||y - z| \, d\mu}.$$

Just because $x - y$, $x - z$, and $y - z$ are members of $L^2[0, 1]$, does this imply that their products are members of $L^1[0, 1]$? Why the square root?

In an admittedly roundabout fashion, we have been led to some *famous inequalities* and *linear spaces*. Here is the most important requirement for a linear space: *Having x, y as members guarantees that $x + y$ and αx are members (for α a scalar).*

As we shall see, there are some inequalities that will be helpful in establishing the triangle inequality for metric spaces.

6.5.2 Famous Inequalities

We begin with Young's Inequality, the work of William Henry Young (1864–1942).

Theorem 6.5.1 (Young, 1912). *For nonnegative numbers a, b,*

$$ab \leq \frac{a^p}{p} + \frac{b^q}{q}, \qquad \text{for } 1 < p < \infty, \ \frac{1}{p} + \frac{1}{q} = 1.$$

Proof. Fix $b > 0$, and maximize the function $f(a) = ab - a^p/p$. $\square$

Next we have the Hölder–Riesz Inequality, which we owe to Otto Hölder (1859–1937) and Frederic Riesz (1880–1956).

Theorem 6.5.2 (Hölder–Riesz, 1889, 1910). *Let $p > 1$ and q satisfy $(1/p) + (1/q) = 1$. If $x \in L^p[0, 1]$ and $y \in L^q[0, 1]$, then $xy \in L^1[0, 1]$ and*

$$L \int_0^1 |xy| \, d\mu \leq \left(L \int_0^1 |x|^p \, d\mu \right)^{1/p} \left(L \int_0^1 |y|^q \, d\mu \right)^{1/q}.$$

Proof. If x or $y \neq 0$ almost everywhere, we have

$$a = \frac{|x|}{\left(L \int_0^1 |x|^p \, d\mu\right)^{1/p}}, \qquad b = \frac{|y|}{\left(L \int_0^1 |y|^q \, d\mu\right)^{1/q}}.$$

Conclude that $xy \in L^1[0, 1]$; integrate; and so on. $\square$

The Minkowski–Riesz Inequality, to which Hermann Minkowski (1864–1909) contributed, is more complex.

Theorem 6.5.3 (Minkowski–Riesz, 1896, 1910). *Let $p \geq 1$. If $x, y \in L^p[0, 1]$, then we have*

$$\left(L \int_0^1 |x + y|^p \, d\mu\right)^{1/p} \leq \left(L \int_0^1 |x|^p \, d\mu\right)^{1/p} + \left(L \int_0^1 |y|^p \, d\mu\right)^{1/p}.$$

Proof. If $p = 1$, integrate the triangle inequality. If $|x + y|^p = 0$ almost everywhere, there is no problem. We will assume $p > 1$ and $\int_0^1 |x + y|^p \, d\mu \neq 0$. Then

$$L \int_0^1 |x + y|^p \, d\mu = L \int_0^1 |x + y|^{p-1} |x + y| \, d\mu$$

$$\leq L \int_0^1 |x + y|^{p-1} |x| \, d\mu + L \int_0^1 |x + y|^{p-1} |y| \, d\mu.$$

Let q satisfy $(1/p) + (1/q) = 1$, that is, $(p - 1)q = p$, and apply the Hölder–Riesz Inequality (Theorem 6.5.2) to the integrals on the right-hand side. Thus,

$$L \int_0^1 |x + y|^{p-1} |x| \, dx \leq \left(L \int_0^1 |x + y|^{(p-1)/q} \, d\mu\right)^{\frac{1}{q}} \left(L \int_0^1 |x|^p \, d\mu\right)^{\frac{1}{p}},$$

and

$$L \int_0^1 |x + y|^{p-1} |y| \, dx \leq \left(L \int_0^1 |x + y|^{(p-1)/q} \, d\mu\right)^{\frac{1}{q}} \left(L \int_0^1 |y|^p \, d\mu\right)^{\frac{1}{p}}.$$

We have

$$L \int_0^1 |x + y|^p \, d\mu \leq \left(L \int_0^1 |x + y|^p \, d\mu\right)^{\frac{1}{q}} \left(L \int_0^1 |x|^p \, d\mu\right)^{\frac{1}{p}}$$

$$+ \left(L \int_0^1 |x + y|^p \, d\mu\right)^{\frac{1}{q}} \left(L \int_0^1 |y|^p \, d\mu\right)^{\frac{1}{p}}.$$

That is,

$$\left(L\int_0^1 |x + y|^p \, d\mu\right)^{1-\frac{1}{q}} \le \left(L\int_0^1 |x|^p \, d\mu\right)^{\frac{1}{p}} + \left(L\int_0^1 |y|^p \, d\mu\right)^{\frac{1}{p}}. \quad \square$$

Exercise 6.5.1. Revisit the examples in Section 6.5.1, and show that the following are metric spaces.

    a. $(C[0, 1], \max_{0 \le t \le 1} |x(t) - y(t)|)$.

    b. $\left(C[0, 1], C\int_0^1 |x(t) - y(t)| \, dt\right)$.

    c. $\left(R[0, 1], R\int_0^1 |x(t) - y(t)| \, dt\right)$.

    d. $\left(L^1[0, 1], L\int_0^1 |x(t) - y(t)| \, dt\right)$.

    e. $\left(L^2[0, 1], \sqrt{L\int_0^1 |x - y|^2 \, d\mu}\right)$.

    Why the square root for $L^2[0, 1]$: $\rho(x, y) = \sqrt{L\int_0^1 |x - y|^2 \, d\mu}$? Among other reasons, if $x = |t| = -y$, $z = 0$, then $L\int_0^1 |x - y|^2 \, d\mu = \frac{4}{3} > L\int_0^1 |x|^2 \, d\mu + L\int_0^1 |y|^2 \, d\mu$.

6.5.3 Completeness

We have some function spaces with metrics. Are these spaces complete? That is, given a Cauchy sequence of elements of these spaces, do we have convergence to an element belonging to the space?

Example 6.5.7. The metric space $(C[0, 1], \max_{0 \le t \le 1} |x(t) - y(t)|)$ is complete.

To explore this example, let's suppose $\{x_n\}$ is a Cauchy sequence of continuous functions on $[0, 1]$. That is, given an $\epsilon > 0$, there is N so that

$$\max_{0 \le t \le 1} |x_n(t) - x_m(t)| < \epsilon \quad \text{for all } n, m \ge N.$$

For each t, $|x_n(t) - x_m(t)| < \epsilon$; in other words, the sequence of real numbers $\{x_n(t)\}$ is a Cauchy sequence. That is, $\lim x_n(t) \equiv x(t)$.

We have a function x on $[0, 1]$. Is x a member of $C[0, 1]$? (Is x continuous on $[0, 1]$?)

Show the convergence is uniform: that $|x_n(t) - x(t)| \leq \epsilon$ for all t in $[0, 1]$ when $n \geq N$. The function x is continuous by Weierstrass's Theorem. Convergence with this metric is uniform convergence.

Note that if $x_n = t^n$ for $0 \leq t \leq 1$, then

$$\lim x_n = \begin{cases} 0 & 0 \leq t < 1, \\ 1 & t = 1. \end{cases}$$

Is $\{x_n\}$ a Cauchy sequence in $(C[0, 1], \max_{0 \leq t \leq 1} |x(t) - y(t)|)$? If so, then by what we have just shown, we have a continuous function x on $[0, 1]$ so that

$$\max_{0 \leq t \leq 1} |x_n(t) - x(t)| \to 0;$$

that is, $x(1) = 1$, and $x = 0$, for $0 \leq t < 1$. We have a contradiction: $\{x_n\}$ is not a Cauchy sequence in $C[0, 1]$ with the metric $\max_{0 \leq t \leq 1} |x - y|$.

Example 6.5.8. The metric space $\left(C[0, 1], C\int_0^1 |x(t) - y(t)| \, dt\right)$ is not complete.

First, let

$$x_n(t) = \begin{cases} n & 0 \leq t \leq 1/n^2, \\ 1/\sqrt{t} & 1/n^2 \leq t \leq 1. \end{cases}$$

Show that $C\int_0^1 |x_n - x_m| \, dt < 2/n$, for $m > n$.

The sequence $\{x_n\}$ is a Cauchy sequence that does not converge to an element of $C[0, 1]$. Claim: No matter what $x \in C[0, 1]$ is chosen, we do not have convergence.

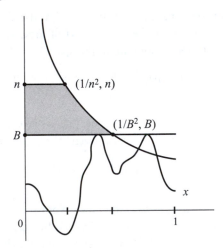

Figure 1. $|x(t) - x_n(t)|$

Hint: Assume such an x. Because x is continuous on $[0, 1]$, x is bounded, $|x| \leq B$, $B \geq 1$. Let $\epsilon > 0$ be given. See Figure 1.

Note that $\text{L}\int_0^1 \left| x_n - (1/\sqrt{t}) \right| \, d\mu = \text{L}\int_0^{1/n^2} n \, d\mu = 1/n \to 0$.

Example 6.5.9. The metric space $\left(R[0, 1], \text{R}\int_0^1 |x(t) - y(t)| \, dt \right)$ is not complete.

The reader can complete the details. Hint: Look at Example 6.5.8. It was critical that x is bounded.

We conclude these explorations on a more positive note.

Example 6.5.10. The metric spaces

$$\left(L^1[0, 1], \text{L}\int_0^1 |x - y| \, d\mu \right) \quad \text{and} \quad \left(L^2[0, 1], \sqrt{\text{L}\int_0^1 |x - y|^2 \, d\mu} \right)$$

are complete — the celebrated Riesz–Fischer Theorem.

The reader should check Example 6.5.8.

6.5.4   The Riesz Completeness Theorem

Frederic Riesz gave us a completeness theorem.

Theorem 6.5.4 (Riesz, 1907). *For $p \geq 1$, the metric space*

$$\left(L^p[0, 1], \left(\text{L}\int_0^1 |x - y|^p \, d\mu \right)^{1/p} \right)$$

is complete.

Proof. We prove the result for $p = 2$. Suppose $\{x_n\}$ is a Cauchy sequence in the metric space $\left(L^2[0, 1], \sqrt{\text{L}\int_0^1 |x - y|^2 \, d\mu} \right)$ and $\epsilon > 0$. We have an N so that

$$\sqrt{\text{L}\int_0^1 |x_n - x_m|^2 \, d\mu} < \epsilon, \qquad \text{for all } n, m \geq N.$$

Choose

$$n_1 \text{ so that } \text{L}\int_0^1 \left| x_n - x_{n_1} \right|^2 \, d\mu < \frac{1}{2}, \qquad n > n_1;$$

$$n_2 > n_1 \text{ so that } \text{L}\int_0^1 \left| x_n - x_{n_2} \right|^2 \, d\mu < \frac{1}{2^2}, \qquad n > n_2; \ldots$$

$$n_k > n_{k-1} \text{ so that } \text{L}\int_0^1 \left| x_n - x_{n_k} \right|^2 \, d\mu < \frac{1}{2^k}, \qquad n > n_k; \ldots$$

In particular,

$$L \int_0^1 |x_{n_2} - x_{n_1}|^2 \, d\mu < \frac{1}{2},$$

$$L \int_0^1 |x_{n_3} - x_{n_2}|^2 \, d\mu < \frac{1}{2^2},$$

$$\cdots$$

$$L \int_0^1 |x_{n_{k-1}} - x_{n_k}|^2 \, d\mu < \frac{1}{2^k}.$$

We are interested in the subsequence $\{x_{n_k}\}$. By the Hölder–Riesz Inequality,

$$L \int_0^1 |x_{n_{k+1}} - x_{n_k}| \, d\mu \le \sqrt{L \int_0^1 (x_{n_{k+1}} - x_{n_k})^2 \, d\mu} \cdot \sqrt{L \int_0^1 1^2 \, d\mu} < \frac{1}{2^k}.$$

So $\sum_{k=1} L \int_0^1 |x_{n_{k+1}} - x_{n_k}| \, d\mu < 1$, and the series converges.

By the Lebesgue Monotone Convergence Theorem, $\left\{ \sum_{k=1}^J |x_{n_{k+1}} - x_{n_k}| \right\}$,

$$\sum \left(L \int_0^1 |x_{n_{k+1}} - x_{n_k}| \, d\mu \right) = L \int_0^1 \sum |x_{n_{k+1}} - x_{n_k}| \, d\mu < 1,$$

and $\sum |x_{n_{k+1}} - x_{n_k}|$ converges almost everywhere. So

$$x_{n_1} + \sum_{k=1}^J \left(x_{n_{k+1}} - x_{n_k} \right) = x_{n_J}$$

converges almost everywhere.

We have $x(t) = \lim x_{n_k}(t)$ when the limit exists and 0 otherwise. Is x a member of $L^2[0, 1] : L \int_0^1 x^2 \, d\mu < \infty$? Apply Fatou's Lemma. Since

$$\lim_j \left(x_{n_j}(t) - x_{n_k}(t) \right)^2 = \left(x(t) - x_{n_k}(t) \right)^2$$

almost everywhere and

$$L \int_0^1 \left(x_{n_j} - x_{n_k} \right)^2 \, d\mu < \frac{1}{2^k}, \qquad \text{for } n_j > n_k,$$

we have

$$L \int_0^1 \left(x - x_{n_k} \right)^2 \, d\mu \le \liminf L \int_0^1 \left(x_{n_j} - x_{n_k} \right)^2 \, d\mu \le \frac{1}{2^k}.$$

So $x - x_{n_k}$ is a member of $L^2[0, 1]$. Since x_{n_k} is also, we may conclude that x is a member of $L^2[0, 1]$.

We have succeeded, therefore, in showing that every Cauchy sequence in the metric space $\left(L^2[0, 1], \sqrt{L \int_0^1 |x - y|^2 \, d\mu} \right)$ contains a subsequence that converges pointwise, almost everywhere, to a member of $L^2[0, 1]$. It remains to show that the original sequence converges to x in the $L^2[0, 1]$ metric.

By the Minkowski Inequality we have

$$\sqrt{L \int_0^1 |x_n - x|^2 \, d\mu} \leq \sqrt{L \int_0^1 |x_n - x_{n_k}|^2 \, d\mu}$$
$$+ \sqrt{L \int_0^1 |x_{n_k} - x|^2 \, d\mu}. \qquad \square$$

Example 6.5.11. Consider the functions $x_1, x_2, \ldots$ illustrated in Figure 2. The sequence $\{x_n\}$ converges at no point of $[0, 1]$. For $p \geq 1$,

$$L \int_0^1 |x_n|^p \, d\mu = \frac{1}{2^{pk}}, \qquad \text{with } 2^k \leq n < 2^{k+1};$$
$$\lim L \int_0^1 |x_n|^p \, d\mu = 0.$$

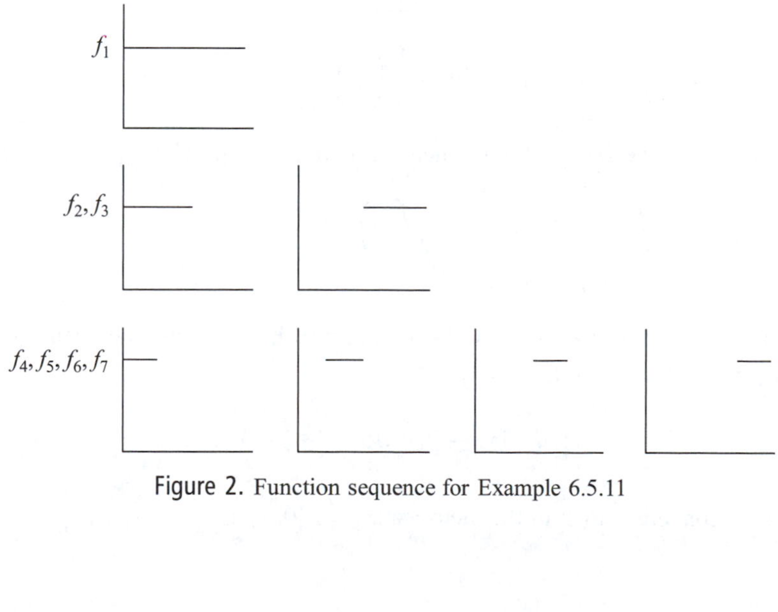

Figure 2. Function sequence for Example 6.5.11

Example 6.5.12. Recall the Cantor set of measure $\frac{1}{2}$ (Section 3.10). Define a sequence of characteristic functions (χ_n) as follows:

$$\chi_1 = \chi_{[0,5/12]} + \chi_{[7/12,1]},$$

$$\chi_2 = \chi_{[0,13/72]} + \chi_{[17/72,5/12]} + \chi_{[7/12,55/72]} + \chi_{[59/72,1]}$$

$$\vdots$$

$$\chi_n = \begin{cases} 1 & \text{on the } 2^n \text{ closed intervals } (F_{n,1}, F_{n,2}, \ldots, F_{n,2^n}), \\ & \text{each of length } 1/2[(1/2^n) + (1/3^n)], \\ 0 & \text{otherwise.} \end{cases}$$

We have

$$\text{R} \int_0^1 \chi_n(t)\, dt = 2^n \cdot \frac{1}{2}\left(\frac{1}{2^n} + \frac{1}{3^n}\right) = \frac{1}{2} + \frac{1}{2}\left(\frac{2}{3}\right)^n,$$

$$\text{R} \int_0^1 [\chi_m(t) - \chi_n(t)]\, dt = \frac{1}{2}\left[\left(\frac{2}{3}\right)^n - \left(\frac{2}{3}\right)^m\right], \qquad \text{for } m > n, \text{ and}$$

$$\lim \chi_n(t) = \chi_C(t),$$

which is the characteristic function on the Cantor set of measure $\frac{1}{2}$.

In this example, we observe first that $\{\chi_n\}$ is a Cauchy sequence in the metric space $\left(L^2[0,1], \sqrt{\text{L}\int_0^1 |x-y|^2\, d\mu}\right)$, such that

$$\sqrt{\text{L}\int_0^1 |\chi_n - \chi_m|^2\, d\mu} < \frac{1}{2}\left(\frac{2}{3}\right)^n, \qquad \text{for } m > n, \text{ and}$$

$$\sqrt{\text{L}\int_0^1 (\chi_n - \chi_C)^2\, d\mu} \leq \frac{1}{2}\left(\frac{2}{3}\right)^n.$$

Next, observe that $\{\chi_n\}$ is a Cauchy sequence in $\left(R[0,1], \text{R}\int_0^1 |x-y|\, dt\right)$:

$$\text{R} \int_0^1 |\chi_n - \chi_m|\, dt < \frac{1}{2}\left(\frac{2}{3}\right)^n, \qquad \text{for } m > n, \text{ and}$$

$$\lim \chi_n(t) \to \chi_C(t),$$

where χ_C is not Riemann integrable (because it is discontinuous on a set of positive measure). However,

$$\text{L} \int_0^1 |\chi_n - \chi_C|\, d\mu = \frac{1}{2}\left(\frac{2}{3}\right)^n.$$

So χ_n converges to χ in the metric space $\left(L[0,1], \text{L}\int_0^1 |x-y|\, dt\right)$.

Exercise 6.5.2.

a. Show the special case of the Hölder–Riesz inequality, $p = 1$, known as Schwarz's Inequality:

$$L \int_0^1 |xy|\, d\mu \le \sqrt{L \int_0^1 x^2\, d\mu} \cdot \sqrt{L \int_0^1 y^2\, d\mu}, \qquad \text{for } x, y \in L^2[0, 1].$$

Hint: $L \int_0^1 (\alpha\,|x| + |y|)^2\, d\mu \ge 0$; also,

$$\left(L \int_0^1 x^2\, d\mu \right) \alpha^2 + 2 \left(L \int_0^1 |xy|\, d\mu \right) \alpha + L \int_0^1 y^2\, d\mu \ge 0.$$

We have a parabola that opens "up" and this has a minimum.

b. Show that in the metric space $\left(L^2[0, 1], \sqrt{L \int_0^1 |x - y|^2\, d\mu} \right)$, limits are unique in the sense that

$$\sqrt{L \int_0^1 |x_n - x|^2\, d\mu} \to 0 \quad \text{and} \quad \sqrt{L \int_0^1 |x_n - y|^2} \to 0$$

imply $x = y$ almost everywhere. Hint: Minkowski Inequality.

c. Show that if $\sqrt{L \int_0^1 |x_n - x|^2\, d\mu} \to 0$, then the sequence $\{x_n\}$ is a Cauchy sequence in $L^2[0, 1]$. Hint: Minkowski: $x_n - x_m$, $x_n - x$, $x_m - x$.

d. Show that if $L \int_0^1 |x_n - x|^2\, d\mu \to 0$, then $\left| L \int_0^1 x_n^2 d\mu - L \int_0^1 x^2 d\mu \right| \to 0$. Hint: By the Minkowski Inequality,

$$\sqrt{L \int_0^1 x_n^2\, d\mu} \le \sqrt{L \int_0^1 (x_n - x)^2\, d\mu} + \sqrt{L \int_0^1 x^2\, d\mu}.$$

Also,

$$\sqrt{L \int_0^1 x_n^2\, d\mu} \le \sqrt{L \int_0^1 (x_n - x)^2\, d\mu} + \sqrt{L \int_0^1 x_n^2\, d\mu},$$

and

$$-\sqrt{L \int_0^1 (x_n - x)^2\, d\mu} \le \sqrt{L \int_0^1 x^2\, d\mu} - \sqrt{L \int_0^1 x_n^2\, d\mu}$$

$$\le \sqrt{L \int_0^1 (x_n - x)^2\, d\mu}.$$

e. Show

$$\left| L\int_0^1 (x_n - x)\, d\mu \right| \le L \int_0^1 |x_n - x|\, d\mu$$

$$\le \sqrt{L\int_0^1 |x_n - x|^2\, d\mu}\,\sqrt{L\int_0^1 1^2\, d\mu}\,.$$

Hint: L^2 convergence implies L^1 convergence on $[0, 1]$; or

$$\lim L\int_0^1 x_n\, d\mu = L\int_0^1 x\, d\mu.$$

6.6 $L^2[-\pi, \pi]$ and Fourier Series

Write an infinite series $\sum c_k$ with the sole requirement that $\sum c_k^2 < \infty$. Now relabel it as $(a_0/\sqrt{2}) + a_1 + b_1 + a_2 + b_2 + \cdots + a_k + b_k + \cdots$, with $(a_0^2/2) + \sum_1 (a_k^2 + b_k^2) < \infty$. This looks suspiciously like a Fourier series (Theorem 2.7.1), doesn't it?

We will construct a function x in $L^2[-\pi, \pi]$ so that its Fourier coefficients are precisely $a_0/2, a_1, b_1, a_2, b_2, \ldots$ and the partial sums converge to x in the metric space

$$\left(L^2[-\pi, \pi],\ \sqrt{L\int_{-\pi}^\pi |x - y|^2\, d\mu} \right).$$

This result is one of the most surprising and astounding in mathematics, but that is the nature of mathematics — beautiful, intriguing:

$$\frac{a_0^2}{2} + \sum_1^\infty (a_k^2 + b_k^2) = \frac{1}{\pi} L\int_{-\pi}^\pi x^2\, d\mu.$$

Now, consider the sequence of $L^2[-\pi, \pi]$ functions

$$\left\{ \frac{a_0}{2} + \sum_1^n (a_k \cos kt + b_k \sin kt) \right\}.$$

How can we show this is a Cauchy sequence in

$$\left(L^2[-\pi, \pi],\ \sqrt{L\int_{-\pi}^\pi |x - y|^2\, d\mu} \right),$$

the given metric space? First, consider that for $m > n$,

$$\int_{-\pi}^{\pi} \left\{ \left[\frac{a_0}{2} + \sum_{1}^{m} (a_k \cos kt + b_k \sin kt) \right] \right.$$

$$\left. - \left[\frac{a_0}{2} + \sum_{1}^{n} (a_k \cos kt + b_k \sin kt) \right] \right\}^2 dt$$

$$= \int_{-\pi}^{\pi} \left(\sum_{n+1}^{m} a_k \cos kt + b_k \sin kt \right)^2 dt$$

$$= \pi \sum_{n+1}^{m} (a_k^2 + b_k^2).$$

Since $(a_0^2/2) + \sum_{1}^{\infty} (a_k^2 + b_k^2)$ converges by assumption,

$$\sum_{n+1}^{m} (a_k^2 + b_k^2) \to 0 \qquad \text{as } n, m \to \infty.$$

By the Riesz–Fischer Theorem (Example 6.5.10), we have a complete metric space. Thus we have a function x belonging to $L^2[-\pi, \pi]$ so that

$$\text{L} \int_{-\pi}^{\pi} \left[\frac{a_0}{2} + \sum_{1}^{n} (a_k \cos kt + b_k \sin kt) - x \right]^2 dt \to 0 \qquad \text{as } n \to \infty.$$

That is, by Exercise 6.5.2,

$$\lim \text{L} \int_{-\pi}^{\pi} \left[\frac{a_0}{2} + \sum_{1}^{n} (a_k \cos kt + b_k \sin kt)^2 \right] dt = \text{L} \int_{-\pi}^{\pi} x^2 \, d\mu,$$

or

$$\frac{a_0^2}{2} + \sum_{1}^{\infty} (a_k^2 + b_k^2) = \frac{1}{\pi} \int_{-\pi}^{\pi} x^2 \, d\mu.$$

Furthermore,

$$\frac{1}{\pi} \text{L} \int_{-\pi}^{\pi} x(t) \cos kt \, dt = a_k, \qquad \text{for } k = 0, 1, 2, \ldots, \text{ and}$$

$$\frac{1}{\pi} \text{L} \int_{-\pi}^{\pi} x(t) \sin kt \, dt = b_k, \qquad \text{for } k = 1, 2, \ldots.$$

For example,

$$\text{L} \int_{-\pi}^{\pi} x(t) \sin mt \, dt$$

$$= \text{L} \int_{-\pi}^{\pi} \left[x(t) - \left(\frac{a_0}{2} + \sum_{1}^{n} a_k \cos kt + b_k \sin kt \right) \right] \sin mt \, dt$$

$$+ \text{L} \int_{-\pi}^{\pi} \left[\frac{a_0}{2} + \sum_{1}^{n} (a_k \cos kt + b_k \sin kt) \right] \sin mt \, dt$$

$$= \text{L} \int_{-\pi}^{\pi} \left[x(t) - \left(\frac{a_0}{2} + \sum_{1}^{n} a_k \cos kt + b_k \sin kt \right) \right] \sin mt \, dt$$

$$+ \begin{cases} 0 & n < m, \\ \pi b_m & n \geq m. \end{cases}$$

As $n \to \infty$, the first term converges to 0:

$$\text{L} \int_{-\pi}^{\pi} \left| x(t) - \left(\frac{a_0}{2} + \sum_{1}^{n} (a_k \cos kt + b_k \sin kt) \right) \sin mt \right| dt$$

$$\leq \sqrt{ \text{L} \int_{-\pi}^{\pi} \left[x(t) - \frac{a_0}{2} + \sum_{1}^{n} (a_k \cos kt + b_k \sin kt) \right]^2 dt }$$

$$\cdot \sqrt{ \text{L} \int_{-\pi}^{\pi} \sin^2 mt \, dt }.$$

Thus we have

$$\text{L} \int_{-\pi}^{\pi} x(t) \sin mt \, dt = \pi b_m \qquad \text{and} \qquad b_m = \frac{1}{\pi} \text{L} \int_{-\pi}^{\pi} x(t) \sin mt \, dt.$$

Exercise 6.6.1. Given $x(t) : \pi - |t|, -\pi \leq t \leq \pi$, consider the Fourier series of x,

$$a_k = \frac{1}{\pi} \text{L} \int_{-\pi}^{\pi} (\pi - |t|) \cos kt \, dt, \qquad \text{for } k = 0, 1, \ldots, \text{ and}$$

$$b_k = \frac{1}{\pi} \text{L} \int_{-\pi}^{\pi} (\pi - |t|) \sin kt \, dt, \qquad \text{for } k = 1, 2, \ldots.$$

a. Show that this Fourier series of x is $\dfrac{\pi}{2} + \dfrac{4}{\pi} \displaystyle\sum_{n=1}^{\infty} \dfrac{\cos(2n - 1)t}{(2n - 1)^2}$.

b. Conclude that $\dfrac{\pi^2}{2} + \dfrac{16}{\pi^2} \displaystyle\sum_{1}^{\infty} \dfrac{1}{(2n-1)^4} = \dfrac{2\pi^2}{3}$. That is,

$$1 + \frac{1}{3^4} + \frac{1}{5^4} + \cdots = \frac{\pi^4}{96} \qquad \text{and}$$

$$1 + \frac{1}{2^4} + \frac{1}{3^4} + \frac{1}{5^4} + \cdots = \frac{\pi^4}{90}.$$

Exercise 6.6.2. Suppose

$$x(t) = \begin{cases} -1 & -\pi < x < 0, \\ 1 & 0 < x < \pi. \end{cases}$$

a. Show that the Fourier series of x is $\dfrac{4}{\pi} \displaystyle\sum \dfrac{1}{2n-1} \sin(2n-1)t$, and evaluate for $t = 1$ and $t = 2$.

b. Conclude that

$$1 + \frac{1}{3^2} + \frac{1}{5^2} + \cdots = \frac{\pi^2}{8} \qquad \text{and}$$

$$1 + \frac{1}{2^2} + \frac{1}{3^2} + \frac{1}{4^2} + \frac{1}{5^2} + \cdots = \frac{\pi^2}{6}.$$

Exercise 6.6.3. Find the Fourier series of

$$x(t) = \begin{cases} (\pi - 1)/2 & 0 < t < 2\pi, \\ 0 & \text{otherwise}, \end{cases}$$

extended periodically. Evaluate at $t = 1$ and $t = 2$.

Beautiful things happen in $L^2[-\pi, \pi]$. In 1966 Lennart Carlson showed that for functions in $L^2[a, b]$ the Fourier series converges *pointwise*, almost everywhere, to the original function.

6.7 Lebesgue Measure in the Plane and Fubini's Theorem

As we constructed Lebesgue measure on R from intervals, we may also construct Lebesgue measure on R^2 from rectangles, on R^3 from...

Such a development may be found many places, but not here. We conclude by stating a result that will be useful later, the theorem of Fubini regarding *double integrals*.

Theorem 6.7.1 (Fubini). *If f is Lebesgue measurable on R^2 and*

$$\int_R \left[\int_R |f(x, y)| \, dx \right] dy \quad \text{or} \quad \int_R \left[\int_R |f(x, y)| \, dy \right] dx$$

exists, then f is Lebesgue integrable on R^2 and

$$\int \int_{R^2} f \, d\mu = \int_R \left[\int_R f(x, y) \, dx \right] dy = \int_R \left[\int_R f(x, y) \, dy \right] dx.$$

Exercise 6.7.1. Given $f(x, y) = (x^2 - y^2)/(x^2 + y^2)$, show

$$\int_0^1 \left(\int_0^1 f(x, y) \, dy \right) dx = \frac{\pi}{4}, \qquad \int_0^1 \left(\int_0^1 f(x, y) \, dx \right) dy = -\frac{\pi}{4}.$$

6.8 Summary

Two Fundamental Theorems of Calculus for the Lebesque Integral

If F is absolutely continuous on $[a, b]$, then

1. F' is Lebesgue integrable on $[a, b]$, and

2. $L \int_a^x F'(t) \, dt = F(x) - F(a)$, with $a \leq x \leq b$.

If f is Lebesgue integrable on $[a, b]$ and $F(x) \equiv L \int_a^x f(t) \, dt$, then

1. F is absolutely continuous on $[a, b]$, and

2. $F' = f$ almost everywhere on $[a, b]$.

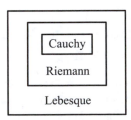

Figure 3. Integrable functions: Cauchy $\subset$ Riemann $\subset$ Lebesgue

6.9 References

1. Apostol, Tom. *Mathematical Analysis*. Reading, Mass.: Addison-Wesley, 1974.

2. Boas, Ralph. *A Primer of Real Functions*. Washington: Mathematical Association of America, 1996.

3. Burk, Frank. *Lebesgue Measure and Integration: An Introduction*. New York: Wiley-Interscience, 1998.

4. Goldberg, Richard. *Methods of Real Analysis*. New York: Wiley, 1964.

5. Gordon, Russell A. *The Integrals of Lebesgue, Denjoy, Perron, and Henstock*. Providence, R.I.: American Mathematical Society, 1994.

6. Kannan, Rangachary, and Carole King Krueger. *Advanced Analysis of the Real Line*. New York: Springer, 1996.

7. Munroe, M. *Measure and Integration*. Reading, Mass.: Addison-Wesley, 1959.

8. Royden, Halsey. *Real Analysis*. New York: Macmillan, 1968.

9. Shilov, G., and B.L. Gurevich. *Integral, Measure and Derivative: A Unified Approach*. New York: Dover, 1977.

10. Spiegel, Murray. *Real Variables*. New York: Schaum's Outline Series 11, 1969.

11. Stromberg, Karl. *An Introduction to Classical Real Analysis*. Belmont, Calif.: Wadsworth, 1981.

12. Titchmarsh, E. *The Theory of Functions*. Oxford University Press, 1939.

CHAPTER 7

The Lebesgue–Stieltjes Integral

Even now there is a very wavering grasp of the true position of mathematics as an element in the history of thought. I will not go so far as to say that to construct a history of thought without profound study of the mathematical ideas of successive epochs is like omitting Hamlet from the play which is named after him. That would be claiming too much. But it is certainly analogous to cutting out the part of Ophelia. This simile is singularly exact. For Ophelia is quite essential to the play, she is very charming — and a little mad. Let us grant that the pursuit of mathematics is a divine madness of the human spirit, a refuge from the goading urgency of contingent happenings.

— Alfred North Whitehead

The Lebesgue measure weights intervals according to their length: $\mu\big((a, b]\big) = b - a$. We are looking for different weightings of intervals, trying to find measures different from Lebesgue measure. Of course the general properties of a measure — nonnegative and countable additivity for disjoint sequences of measurable sets — must be retained. We will discuss three of the most common approaches, which differ from each other essentially in their starting points.

7.1   L-S Measures and Monotone Increasing Functions

In Chapter 4 we explored the particularly fruitful, and most fundamental, approach of Thomas Stieltjes, who modelled mass distributions with monotone increasing right-continuous functions. His approach involved the variation of ϕ on an interval $(a, b]$: $\phi(b) - \phi(a)$.

For Lebesgue–Stieltjes measure, we may also begin with the variation of a monotone increasing right-continuous F on the half-open interval $(a, b]$

as $F(b) - F(a)$, in other words the entity ("total mass" to b minus "total mass" to a), but we now proceed to develop a measure.

Whereas Lebesgue measure weights the interval $(a, b]$ by $b - a$, Lebesgue–Stieltjes measure will weight $(a, b]$ by $F(b) - F(a)$. We have a weight function τ: $\tau((a, b]) = F(b) - F(a)$ and $\tau(\phi) = 0$. Since familiarity with the weight function τ will be helpful, let's examine its properties.

7.1.1   Properties of the Weight Function

As an exercise, demonstrate each of these properties. Some suggestions are provided.

Property 1. For $a < c < b$,

$$\tau((a, b]) = F(b) - F(a)$$
$$= F(b) - F(c) + F(c) - F(a)$$
$$= \tau((c, b]) + \tau((a, c]).$$

Property 2. If $(a_k, b_k]$, $1 \leq k \leq n$, are disjoint and $\cup_1^n (a_k, b_k] \subset (a, b]$, then $\sum_1^n \tau((a_k, b_k]) \leq \tau((a, b])$. Hint: Relabel so $a \leq a_1 < b_1 \leq a_2 < b_2 \leq \cdots \leq a_n < b_n \leq b$. Then

$$\tau((a, b]) = F(b) - F(a) \geq F(b_n) - F(a_1)$$
$$\geq \sum_1^n [F(b_k) - F(a_k)]$$
$$= \sum_1^n \tau((a_k, b_k]).$$

Property 3. If $(a, b] \subset \cup_1^n (a_k, b_k]$, then $\tau((a, b]) \leq \sum_1^n \tau((a_k, b_k])$. Hint: Assume all intervals are needed and $a_1 \leq a_2 \leq \cdots \leq a_n$. Then $a_1 \leq a \leq b_1$, $a_n < b \leq b_n$, with $a_{k+1} \leq b_k < b_{k+1}$, and

$$\tau((a, b]) = F(b) - F(a)$$
$$\leq \sum_1^{n-1} [F(a_{k+1}) - F(a_k)] + F(b_n) - F(a_n)$$
$$\leq \sum_1^n \tau((a_k, b_k]).$$

Property 4. If $(a_k, b_k]$, $1 \leq k \leq n$, are disjoint and $(a, b] = \cup_1^n (a_k, b_k]$, then $\tau((a, b]) = \sum_1^n \tau((a_k, b_k])$.

Property 5. If $(a, b] \subset \cup_1^\infty (a_k, b_k]$, then $F(b) - F(a) \leq \sum_1^\infty \tau((a_k, b_k])$.
Hint:

a. $[a + \delta, b] \subset \cup_1^\infty (a_k, b_k] \subset \cup_1^\infty (a_k, b_k^*)$, where $b_k < b_k^*$, and $F(a + \delta) < F(a) + \epsilon$.

b. $(a, b] = (a, a + \delta] \cup [a + \delta, b] \subset (a, a + \delta] \cup$
$\left(\cup_{i=1}^N (a_{k_i}, b_{k_i}^*) \subset (a, a + \delta] \right) \cup \left(\cup_{i=1}^N (a_{k_i}, b_{k_i}^*] \right)$. (Heine–Borel).

c. Using Property 3,

$$\tau((a, b]) \leq F(a + \delta) - F(a) + \sum_1^N \left[F(b_{k_i}^*) - F(a_{k_i}) \right]$$

$$< \epsilon + \sum_1^N \left[F(b_{k_i}) - F(a_{k-i}) + \frac{\epsilon}{2^{k_i}} \right]$$

$$\leq \sum_1^\infty \tau((a_k, b_k]) + 2\epsilon.$$

d. $F(b) - F(a) - 2\epsilon < \sum_1^\infty \tau((a_k, b_k])$. By the arbitrary nature of ϵ, the conclusion follows.

Property 6. If $(a, b] = \cup_1^\infty (a_k, b_k]$, with mutually disjoint intervals, then $\tau((a, b]) = \sum_1^\infty \tau((a_k, b_k])$. Hint: Since $(a, b] \subseteq \cup(a_k, b_k]$, by Property 5, $\tau((a, b]) \leq \sum \tau((a_k, b_k])$.
For the other direction, $\cup_1^n (a_k, b_k] \subset (a, b]$. By Property 3,

$$\sum_1^n \tau((a_k, b_k]) \leq \tau((a, b]),$$

independent of n. Thus

$$\sum_1^\infty \tau((a_k, b_k]) \leq \tau((a, b]).$$

So τ is countably additive on half-open, half-closed subintervals.

7.1.2   The L-S Outer Measure

Knowing how τ operates, as the analog of length ℓ in Lebesgue measure, we define the *outer measure*.

Definition 7.1.1 (Lebesgue–Stieltjes Outer Measure). The Lebesgue–Stieltjes outer measure μ_F^*, determined by a monotone increasing right-continuous function F, for any set A of real numbers, is defined as

$$\mu_F^*(A) = \inf\left\{\sum_1^\infty \tau\big((a_k, b_k]\big) \mid A \subset \cup_1^\infty(a_k, b_k]\right\},$$

with $\tau\big((a_k, b_k]\big) = F(b_k) - F(a_k)$ and $\tau(\phi) = 0$.

Exercise 7.1.1. Show that μ_F^* satisfies the general requirements for an outer measure.

   a. $\mu_F^*(\phi) = 0$.

   b. If $A \subset B$, then $\mu_F^*(A) \le \mu_F^*(B)$ (monotonicity).

   c. $\mu_F^*(\cup A_k) \le \sum \mu_F^*(A_k)$ (subadditivity). Hint: Review Section 5.2.1, the Lebesgue Wish List.

7.2   Carathéodory's Measurability Criterion

As with Lebesgue measure, from Lebesgue–Stieltjes outer measure we set a measurability requirement. This again is the work of Constantin Carathéodory (1873–1950).

Definition 7.2.1 (Carathéodory's Measurability Criterion). A set of real numbers E will be Lebesgue–Stieltjes measurable if

$$\mu_F^*(A) = \mu_F^*(A \cap E) + \mu_F^*(A \cap E^c)$$

holds for every set of real numbers A.

   The next series of exercises demonstrates that intervals are μ_F measurable.

Exercise 7.2.1.

   a. Show that $\mu_F^*\big((a, b]\big) = F(b) - F(a) = \tau\big((a, b]\big)$. Hint: $(a, b] \subset (a, b]$, so $\mu_F^*\big((a, b]\big) \le \tau\big((a, b]\big) = F(b) - F(a)$. For the other direction, use the properties in Section 7.1.1.

   b. Show that intervals of the form $(-\infty, b]$ satisfy Carathéodory's Criterion, $\mu_F^*(A) = \mu_F^*\big(A \cap (-\infty, b]\big) + \mu_F^*\big(A \cap (b, \infty)\big)$. Hint: Section 5.3.1.

c. Show that intervals are μ_F measurable.

We are mimicking our development of Lebesgue measure in Chapter 5. Along those lines, μ_F^* measurability of intervals implies that Borel sets are likewise Lebesgue–Stieltjes measurable. Indeed, we have a measure space: $(R, \mathcal{B}, \mu_F)$. A monotone increasing right-continuous F determines a measure (Lebesgue–Stieltjes) μ_F on a sigma algebra of sets that contains the Borel sets, the Borel sigma algebra $\mathcal{B}$.

Here are some examples of measuring with μ_F. In the first example, notice that if F is continuous everywhere, then "points" have Lebesgue–Stieltjes measure zero.

Example 7.2.1.

$$\mu_F(\{a\}) = \mu_F\left(\cap\left(a - \frac{1}{n}, a + \frac{1}{n}\right]\right)$$

$$= \lim \mu_F\left(\left(a - \frac{1}{n}, a + \frac{1}{n}\right]\right)$$

$$= F(a) - F(a^-).$$

Example 7.2.2.

a. $\mu_F\left((-\infty, x]\right) = \mu_F\left(\cup(x - k, x]\right) = \lim \mu_F\left((x - k, x]\right)$

$\qquad\qquad = F(x) - \lim_{x \to -\infty} F(x) = F(x) - F(-\infty).$

b. $\mu_F\left((-\infty, x)\right) = F(x^-) - F(-\infty).$

c. $\mu_F\left((x, \infty)\right) = F(\infty) - F(x).$

d. $\mu_F\left([x, \infty)\right) = F(\infty) - F(x^-).$

e. $\mu_F(R) = F(\infty) - F(-\infty).$

Example 7.2.3.

a. $\mu_F\left([a, b]\right) = \mu_F(\{a\}) + \mu_F\left((a, b]\right) = F(b) - F(a^-).$

b. $\mu_F\left([a, b)\right) = F(b^-) - F(a^-).$

c. $\mu_F\left((a, b)\right) = F(b^-) - F(a).$

Exercise 7.2.2.

a. Calculate $\mu_F\left((-\infty, \infty)\right)$, $\mu_F\left((-1, 1]\right)$, and $\mu_F(\{0\})$, given

$$F(x) = \begin{cases} x & x < 0, \\ x + 1 & x \geq 0. \end{cases}$$

b. Calculate $\mu_F(\{\text{rationals in } (0, 1)\})$, $\mu_F = (\{\text{irrationals in } (0, 1)\})$, and $\mu_F(C)$, where C is the Cantor set (Section 3.9), given

$$F(x) = \begin{cases} 0 & x \le 0, \\ 2x & 0 \le x \le 1, \\ 2 & x \ge 1. \end{cases}$$

c. Calculate $\mu_F([0, 1])$ and $\mu_F(C)$, given

$$F(x) = \begin{cases} 0 & x \le 0, \\ \text{Cantor function} & 0 \le x \le 1, \\ 1 & x \ge 1. \end{cases}$$

d. Calculate $\mu_F((a, b])$, given

$$F(x) = \begin{cases} 0 & x < 0, \\ 1 & x \ge 0. \end{cases}$$

7.3 Avoiding Complacency

For many years I took for granted Stieltjes' contributions in the setting of Lebesgue–Stieltjes measure. Recently I read a beautiful book, *Measure, Topology, and Fractal Geometry*, where I came across the following example (Edgar, 1990, p. 136). I have modified it slightly for our purposes.

Suppose we weight the half-open interval $(a, b]$ by $\tau((a, b]) = \sqrt{b - a}$. We then have an outer measure μ_τ^* that will determine a sigma algebra of measurable sets, and thus we have a measure μ_τ. No problem there.

But what is the measure of the interval $[-1, 0]$? It turns out that this interval is not μ_τ measurable.

Certainly $\mu_\tau^*((0, 1]) \le 1$, since $(0, 1]$ is a cover of itself. On the other hand, if $(0, 1] \subseteq \cup(a_k, b_k]$, then

$$\left(\sum \sqrt{b_k - a_k} \right)^2 = \sum \left(\sqrt{b_k - a_k} \right)^2 + 2 \sum_{i \ne j} \sqrt{b_j - a_j} \sqrt{b_i - a_i}$$

$$\ge \sum (b_k - a_k) \ge 1.$$

Thus $1 \le \sum \sqrt{b_k - a_k}$. By the infimum property, $1 \le \mu_\tau^*((0, 1])$. So $\mu_\tau^*((0, 1]) = 1$.

Show that $\mu_\tau^*((-1, 0]) = 1$.

For the interval $(-1, 1]$, $\mu_\tau^*((-1, 1]) \leq \sqrt{2}$. By Carathéodory's Criterion, if $[-1, 0]$ is μ_τ measurable, then

$$\mu_\tau^*(A) = \mu_\tau^*(A \cap [-1, 0]) + \mu_\tau^*(A \cap [-1, 0]^c)$$

for all sets A of real numbers. In particular, if $A = (-1, 1]$, then

$$\sqrt{2} \geq \mu_\tau^*((-1, 1]) = \mu_\tau^*((-1, 1] \cap [-1, 0]) + \mu_\tau^*((-1, 1] - [-1, 0])$$
$$= \mu_\tau^*((-1, 0]) + \mu_\tau^*((0, 1]) = 2.$$

Of course in hindsight, we can see that we lacked finite additivity, that is, $\sqrt{b - a} \neq \sqrt{b - a} + \sqrt{c - a}$, for $a < c < b$ — so we might have been suspicious.

7.4   L-S Measures and Nonnegative Lebesgue Integrable Functions

Suppose f is a nonnegative Lebesgue integrable function on the reals. Thus, f is nonnegative, Lebesgue measurable, and $\int_R f\, d\mu < \infty$. Now define a function F on the reals by $F(x) = \int_{-\infty}^x f\, d\mu$. The function F is clearly nondecreasing, bounded, and absolutely continuous (by Theorem 6.4.1).

As before, we may construct a Lebesgue–Stieltjes measure μ_f, and we have a sigma algebra that includes the Borel sets. Even nicer, the sigma algebra of μ_f measurable sets contains the Lebesgue measurable sets. Let's look at some properties of this measure.

7.4.1   Properties of μ_f

Assuming that $F(x) = \int_{-\infty}^x f\, d\mu$ and that f is nonnegative and Lebesgue integrable, we can show that Lebesgue measurable sets are μ_f measurable sets. In fact, $\mu_f(E) = L \int_E f\, d\mu$ when E is a Lebesgue measurable subset of R.

Property 1. If $\mu(E) = 0$, then E is μ_f measurable and $\mu_f(E) = 0$.

Let $\epsilon > 0$ be given. Because F is absolutely continuous we have a δ so that

$$\mu\left(\cup_{k=1}^n (a_k, b_k]\right) < \delta \quad \text{implies} \quad \sum [F(b_k) - F(a_k)] < \epsilon.$$

Because $\mu(E) = 0$, we have a cover of E by disjoint half-open intervals $(a_k, b_k]$ so that $\mu\left(\cup_1^\infty (a_k, b_k]\right) < \delta$. But then, $\mu\left(\cup_1^n (a_k, b_k]\right) < \delta$ for all

n. Thus

$$\mu_f\left(\cup_1^n(a_k,b_k]\right) = \sum_1^n [F(b_k) - F(a_k)] < \epsilon,$$

independent of n. We have $\mu_f^*(E) \le \mu_f^*\left(\cup_1^\infty(a_k,b_k]\right) \le \epsilon$. That is, $\mu_f^*(E) = 0$. But sets of Lebesgue–Stieltjes outer measure zero are Lebesgue–Stieltjes measurable:

$$0 \le \mu_f^*(A \cap E) \le \mu_f^*(E) = 0, \qquad \text{and}$$
$$\mu_f^*(A) \ge \mu_f^*(A \cap E^c). \qquad \text{(monotonicity)}$$

Property 2. If the set E is Lebesgue measurable, then E is Lebesgue–Stieltjes measurable (μ_f).

From Exercise 5.6.1, every Lebesgue measurable set (E) may be written as the union of a Borel set (B) and a Lebesgue measurable set of measure zero ($E - B$). That is, $E = B \cup (E - B)$.

By Property 1, $E - B$ is μ_f measurable. Borel sets are μ_f measurable (Exercise 7.2.1). Thus E is μ_f measurable and

$$\mu_f(E) = \mu_f(B) + \mu_f(E - B) = \mu_f(B).$$

The sigma algebra of μ_f measurable sets contains $\mathcal{M}$, the sigma algebra of Lebesgue measurable sets. We have a measure space: $(R, \mathcal{M}, \mu_f)$.

Property 3. With F continuous,

$$\mu_f\big((a,b]\big) = \mu_f\big((a,b)\big) = \mu_f\big([a,b)\big) = \mu_f\big([a,b]\big)$$
$$= L\int_a^b f \, d\mu$$
$$= \int_{(a,b]} f \, d\mu.$$

Example 7.4.1. For any nonnegative Lebesgue integrable function f, define a nonnegative set function, $\hat{\mu}$, on the Lebesgue measurable sets E by $\hat{\mu}(E) = L\int_E f \, d\mu$. Will $\hat{\mu}$ be a measure?

We will explore this question in stages.

a. If $A \subset B$, show $\hat{\mu}(A) \le \hat{\mu}(B)$.

b. If $\{E_k\}$ is a mutually disjoint sequence of Lebesgue measurable sets, show that $\hat{\mu}(\cup E_k) = \sum \hat{\mu}(E_k)$. Hint: Monotone convergence (Theorem 6.3.2) and countable additivity. We have a measure on Lebesgue measurable sets: μ_f.

c. Note that

$$\hat{\mu}_f((a,b]) = L\int_{-\infty}^{b} f\, d\mu - L\int_{-\infty}^{a} f\, d\mu$$
$$= F(b) - F(a) = \mu_F((a,b]).$$

d. Given that E is Lebesgue measurable, does $\hat{\mu}_f(E) = \mu_f(E)$? Because $\hat{\mu}_f$ and μ_f agree on the covering sets, $(a,b]$, and because R can be written as a countable union of pairwise disjoint sets, $R = \cup(n, n+1]$, it can be shown that we have uniqueness: $\hat{\mu}_f(E) = \mu_f(E)$, E Lebesgue measurable.

Exercise 7.4.1.

a. Calculate $\mu_f(R)$, $\mu_f([-1,1])$, $\mu_f((-1,1])$, $\mu_f(\{0\})$, and $\mu(\{0\})$, given

$$f(x) = e^{-|x|}, \qquad F(x) = L\int_{-\infty}^{x} f\, d\mu, \quad -\infty < x < +\infty.$$

b. Calculate $\mu_f((-1,1))$, $\mu_f((-1,1])$, and $\mu_f((-1,2))$, and compare with Exercise 7.2.2, given

$$f(x) = \begin{cases} 0 & x < 0, \\ 2 & 0 \le x \le 1, \\ 0 & x > 1, \end{cases} \qquad F(x) = L\int_{-\infty}^{x} f\, d\mu.$$

c. Calculate $\mu_f((-1,1])$ and compare with $\mu((-1,1])$, given

$$f(x) = \begin{cases} 0 & |x| > 1, \\ 2x + 2 & -1 \le x < 0, \\ -2x + 2 & 0 \le x \le 1, \end{cases} \qquad F(x) = L\int_{-\infty}^{x} f\, d\mu.$$

d. Given the Gaussian probability density function,

$$f(x) = \frac{1}{\sqrt{2\pi t}} e^{-x^2/2t}, \quad t > 0, \quad \text{and} \quad F(x) \equiv L\int_{-\infty}^{x} f\, d\mu,$$

show $\lim_{x\to\infty} F(x) = 1$. Approximate $\mu_f((-1,1])$. Calculate $\lim_{x\to-\infty} F(x)$.

e. Calculate F given the Cauchy density function,

$$f(x) = \frac{1}{\pi} \frac{1}{1 + x^2}, \quad \text{and} \quad F(x) \equiv L\int_{-\infty}^{x} f\, d\mu.$$

7.5   L-S Measures and Random Variables

Our final approach begins with a probability space Ω, a sigma algebra of subsets of this space Σ, and a probability measure $P: (\Omega, \Sigma, P)$.

Let X be a random variable on this space. That is, X is a real-valued function defined on Ω, and the inverse image of the interval $(-\infty, x]$, $X^{-1}((-\infty, x])$, is a member of the sigma algebra Σ for every real number x. Borel sets are P measurable.

We may calculate the measure of this inverse image: $P\left(X^{-1}((-\infty, x])\right)$. We define an extended real-valued function F on the reals by

$$F_X(x) = P\left(X^{-1}((-\infty, x])\right),$$

the probability distribution function $F: R \to [0, 1]$.

7.5.1   Properties of F_X

It will help to consider some characteristics of F_X.

Property 1. The function F_X is monotone increasing on R:

$$P\left(X^{-1}((-\infty, x])\right) \le P\left(X^{-1}((-\infty, y])\right), \qquad x \le y.$$

Property 2. The function F_X is right-continuous on R:

$$(-\infty, x] = \cap\left(-\infty, x + \frac{1}{n}\right],$$

$$X^{-1}((-\infty, x]) = \cap X^{-1}\left(\left(-\infty, x + \frac{1}{n}\right]\right),$$

P is a measure, and so on.

Property 3. The limit $\lim_{x \to -\infty} F_X(x) = 0$, since $\phi = \cap(-\infty, -n]$: $F_X(-\infty) = 0$.

Property 4. The limit $\lim_{x \to +\infty} F_X(x) = 1$ since $R = \cup(-\infty, n]$: $F_X(\infty) = 1$.

Again, we have a monotone increasing right-continuous function F_X on the reals, and thus we may construct a Lebesgue–Stieltjes measure, μ_X, with the sigma algebra of μ_X measurable sets containing the sigma algebra of Borel sets $\mathcal{B}$. The probability space $(R, \mathcal{B}, \mu_X)$ is said to be *induced* by the random variable X.

This concludes our treatment of generating Lebesgue–Stieltjes measures. Now our development of the Lebesgue–Stieltjes integral will proceed smoothly.

7.6   The Lebesgue–Stieltjes Integral

To construct the Lebesgue–Stieltjes Integral, we will mimic the construction of the Lebesgue integral (Section 6.2.4), with μ_S denoting μ_F, μ_f, or μ_X respectively. We proceed in four stages.

Step 1. A real-valued function g is said to be μ_S measurable if $g^{-1}((-\infty, x])$ — or equivalently $g^{-1}(B)$, B a Borel set — belongs to the appropriate sigma algebra. Since the sigma algebras for μ_F, μ_f, or μ_X contain, respectively, $\mathcal{B}$, $\mathcal{M}$, or $\mathcal{B}$, we will assume $g^{-1}(B)$ belongs to $\mathcal{B}$, $\mathcal{M}$, or $\mathcal{B}$, in that order.

Step 2. For nonnegative simple functions ϕ, $\phi = \sum_{k=1}^{n} c_k \chi_{E_k}$, where $E = \cup E_k$, with E_k mutually disjoint, μ_S measurable sets, and c_k nonnegative real numbers, we define the Lebesgue–Stieltjes integral of ϕ by

$$\text{L-S} \int_E \phi \, d\mu_S = \sum_{k=1}^{n} c_k \mu_S(E_k).$$

Step 3. If g is a nonnegative μ_S measurable function defined on a μ_S measurable set E, we may approximate g by a monotonically increasing sequence of simple functions $\{\phi_k\}$. Thus, the sequence $\{\text{L-S} \int_E \phi_k \, d\mu_S\}$ is monotone increasing and has a limit (extended reals, perhaps). We define

$$\text{L-S} \int_E g \, d\mu_S \equiv \lim \text{L-S} \int_E \phi_k \, d\mu_S.$$

Step 4. If g is a μ_S measurable function, then

$$g = \frac{|g| + g}{2} - \frac{|g| - g}{2},$$

and we have monotone increasing sequences of simple functions ϕ_k^+ and ϕ_k^- converging respectively to $(|g| + g)/2$ and $(|g| - g)/2$. We define

$$\text{L-S} \int_E g \, d\mu_S = \lim \text{L-S} \int_E \phi_k^+ \, d\mu_S - \lim \text{L-S} \int_E \phi_k^- \, d\mu_S,$$

provided both integrals on the right are finite.

Exercise 7.6.1.

a. Calculate L-S $\int_{(-1,1]} g \, d\mu_F$, given

$$F(x) = \begin{cases} -1 & x < 0, \\ 2 & x \geq 0, \end{cases} \quad \text{and} \quad g(x) = 2, \quad -1 \leq x \leq 1.$$

b. Calculate L-S $\int_{(-1,1]} g \, d\mu_F$, given

$$F(x) = \begin{cases} x - 1 & x < 0, \\ 2x + 2 & x \geq 0, \end{cases} \quad \text{and} \quad g(x) = 2, \quad -1 \leq x \leq 1.$$

c. Calculate L-S $\int_{(-1,1]} g \, d\mu_F$, given

$$F(x) = \begin{cases} -1 & x < 0, \\ 2 & x \geq 0, \end{cases} \quad \text{and} \quad g(x) = 2 + x^2, \quad -1 \leq x \leq 1.$$

d. Calculate L-S $\int_{(-1,2]} x^2 \, d\mu_f$, given

$$f(x) = \begin{cases} 0 & x \leq 0, \\ 2 & 0 < x < 1, \\ 0 & x \geq 1. \end{cases}$$

e. Determine $F_X(x)$ and calculate L-S $\int_{(-1,2]} x \, d\mu_X$, given $(\Omega, \Sigma, P) = (R, \mathcal{M}, P)$, with $\mathcal{M}$ a sigma algebra of Lebesgue measurable sets, P a probability measure, $P(E) \equiv L \int_E \chi_{[0,1]} \, d\mu$, and

$$X(\omega) = \begin{cases} 0 & \omega \leq 0, \\ 2\omega & 0 < \omega \leq 1/2, \\ 2 - 2\omega & 1/2 < \omega \leq 1, \\ 0 & 1 < \omega. \end{cases}$$

7.7   A Fundamental Theorem for L-S Integrals

We conclude this chapter with a Fundamental Theorem of Calculus for Lebesgue–Stieltjes integrals.

Theorem 7.7.1 (FTC for Lebesgue–Stieltjes Integrals). *If g is a Lebesgue measurable function on R, f is a nonnegative Lebesgue integrable function on R, and $F(x) = L \int_{-\infty}^{x} f \, d\mu$, then:*

1. *F is bounded, monotone increasing, absolutely continuous, and differentiable almost everywhere, and $F' = f$ almost everywhere.*

2. *We have a Lebesgue–Stieltjes measure μ_f so that, for any Lebesgue measurable set E, $\mu_f(E) = L \int_E f \, d\mu$, and μ_f is absolutely continuous with respect to Lebesgue measure.*

3. L-S $\int_R g \, d\mu_f = L \int_R g f \, d\mu = L \int_R g F' \, d\mu.$

Proof. The first two parts of the conclusion have already been discussed among the properties of μ_f (Section 7.4.1). As for the last conclusion, we will give only a sketch.

Step 1. Consider $g = \chi_E$, for E a Lebesgue measurable set. We have

$$\text{L-S}\int_R \chi_E \, d\mu_f = \text{L-S}\int_E d\mu_f = \mu_f(E) = \text{L}\int_E f \, d\mu$$

$$= \text{L}\int_R \chi_E f \, d\mu = \text{L}\int_R \chi_E F' \, d\mu.$$

Step 2. For the simple function ϕ, with $c_k \geq 0$, we have $\phi = \sum_1^n c_k \chi_{E_k}$. No problem; linearity of the integral.

Step 3. For the nonnegative simple function ϕ_k, with $0 \leq \phi_k \leq \phi_{k+1}$, we have $g = \lim \phi_k$. Monotone Convergence Theorem... and so on. $\quad\square$

This concludes our treatment of the Lebesgue–Stieltjes integral.

7.8 Reference

1. Edgar, Gerald. *Measure, Topology and Fractal Geometry.* New York: Springer-Verlag, 1990.

CHAPTER 8

The Henstock–Kurzweil Integral

He who knows not mathematics and the results of recent scientific investigation dies without knowing truth. — K. H. Schellbach

In this chapter we present a beautiful extension of the Lebesgue integral obtained by an apparently slight modification of the Riemann integration process. Recall that in Section 3.12 we saw functions with a bounded derivative whose derivative was not Riemann integrable. These examples prompted Lebesgue to develop an integration process by which differentiable functions with bounded derivatives could be reconstructed from their derivatives:

$$\text{L} \int_a^x F' \, d\mu = F(x) - F(a).$$

This was a Fundamental Theorem of Calculus for the Lebesgue integral (Theorem 6.4.2).

The next step would be to try to remove the "bounded" requirement on the derivative. We want an integration process in which all Lebesgue integrable functions will still be integrable and where differentiability of F guarantees $\int_a^x F'(t) \, dt = F(x) - F(a)$. Denjoy (in 1912) and Perron (in 1914) successfully developed such extensions: see Gordon's book, *The Integrals of Lebesgue, Denjoy, Perron, and Henstock* (1994).

In 1957, Jaroslav Kurzweil utilized a generalized version of the Riemann integral while studying differential equations. Independently, Ralph Henstock (1961) discovered and made a comprehensive study of this generalized Riemann integral, which we will call the *Henstock–Kurzweil integral* or H-K integral. In Gordon's wonderful book all these integrals are fully developed and shown to be equivalent. We will use the constructive approach due to Kurzweil and Henstock because of its relative simplicity.

8.1   The Generalized Riemann Integral

Recall the Riemann integration process for a bounded function f on the interval $[a, b]$:

1. Divide $[a, b]$ into a finite number of contiguous intervals.

2. Select a tag c_k in each subinterval $[x_{k-1}, x_k]$ at which to evaluate f.

3. Form the collection of point intervals consisting of $(c_1, [x_0, x_1])$, $(c_2, [x_1, x_2])$, ..., $(c_n, [x_{n-1}, x_n])$.

4. Calculate the associated Riemann sum, $\sum_{k=1}^{n} f(c_k)(x_k - x_{k-1})$.

If we find that these Riemann sums — these numbers — are close to a number A for all collections of point intervals with subintervals of a uniformly small length $(x_k - x_{k-1}) < \delta$, with δ constant, then we declare f to be Riemann integrable on $[a, b]$ and write $R \int_a^b f(x)\, dx = A$.

In the modified Riemann integration process that we are about to describe, the local behavior of f plays a prominent role. Therefore, instead of dividing $[a, b]$ into a finite number of contiguous intervals of uniform length and then selecting the tag c in each subinterval at which to evaluate f, we will first examine the points where f is not well behaved. For instance, we will look for jumps or rapid oscillation. Using that information, we will divide $[a, b]$ into contiguous subintervals of variable length. In this way, we will be able to control the erratic behavior of f about a tag c by controlling the size (the length) of the associated subinterval $[u, v]$, where $u \leq c \leq v$.

Generally, if f does not change much about the point c, the length of the associated subinterval $[u, v]$ may be large. But if f behaves erratically (jumps, oscillates, etc.) about c, then the length of the associated subinterval $[u, v]$ needs to be small.

This is the key idea: We want a function, a positive function $\delta(\cdot)$, on $[a, b]$, that associates small intervals $[u, v]$ about c when f exhibits erratic behavior at c. In particular, $c - \delta(c) < u \leq c \leq v < c + \delta(c)$:

$$[u, v] \subset \big(c - \delta(c), c + \delta(c)\big) \qquad \text{and} \qquad v - u < 2\delta(c).$$

The term in the Riemann sum, $f(c)(v - u)$, would be dominated by $f(c) \cdot 2\delta(c)$.

We will clarify this process with examples. As always, we are driven by considerations of area, and thus begin with examples whose integrals, because of "areas," clearly "should be..."

Example 8.1.1. Suppose

$$f(x) = \begin{cases} 2 & x \neq 1, \\ 100 & x = 1. \end{cases}$$

From area considerations, the H-K integral of f over the interval $[0, 3]$ should be 6:

$$\text{H-K} \int_0^3 f(x)\, dx = 6.$$

Let's see why this is so. For any ordinary Riemann partition of $[0, 3]$ with all the tags c_k different from $x = 1$,

$$\sum f(c_k)(x_k - x_{k-1}) = \sum 2(x_k - x_{k-1}) = 2 \cdot 3 = 6.$$

However, $x = 1$ may be the tag of one interval, say $(1, [x_{k-1}, x_k])$, or it may be the tag of two adjacent subintervals, $(1, [x_{k-1}, 1])$, $(1, [1, x_{k+1}])$. The difference of such a sum and the number 6 is bounded in absolute value by $|f(1) - 2|\,(x_k - x_{k-1})$, $|f(1) - 2|\,(x_{k+1} - x_{k-1})$, respectively.

So, given $\epsilon > 0$, the difference between any ordinary Riemann sum and the number 6 will be bounded by ϵ if the length of the subinterval(s) containing the tag $x = 1$, $x_k - x_{k-1}(x_k - x_{k-1}, x_{k+1} - x_k)$, is less than $\epsilon / \{4[\,|f(1)| + 2]\}$:

$$|f(1) - 2|\,(x_k - x_{k-1}) < \left[\,|f(1)| + 2\right] \cdot \frac{\epsilon}{4[\,|f(1)| + 2]} = \frac{\epsilon}{4},$$

or

$$|f(1) - 2|\,(x_{k+1} - x_{k-1})$$
$$< \left(\left[\,|f(1)| + 2\right] \cdot \frac{\epsilon}{4[\,|f(1)| + 2]}\right) + \left(\left[\,|f(1)| + 2\right] \cdot \frac{\epsilon}{4[\,|f(1)| + 2]}\right)$$
$$= \frac{\epsilon}{2}.$$

If we can guarantee that for any partition of $[a, b]$ the subinterval containing $x = 1$ has length less than $\epsilon / \{4[\,|f(1)| + 2]\}$, then we will conclude that the integral has value 6. Obviously we could require all the subintervals to have a length less than $\epsilon / \{4[\,|f(1)| + 2]\}$; and this is the usual Riemann integral when δ is a constant, in this case, $\epsilon / \{4[\,|f(1)| + 2]\}$.

As we have discussed however, the only subintervals of importance are those that contain the point of discontinuity, $x = 1$. Thinking of δ as a function, we have

$$\delta(x) \equiv \begin{cases} 1 & x \neq 1, \\ \epsilon / \{4[|f(1)| + 2]\} & x = 1. \end{cases}$$

If this function $\delta(\cdot)$ forces the subinterval of any partition that contains $x = 1$ to have length less than $\epsilon / \{4[|f(1)| + 2]\}$, the integral would have value 6.

Exercise 8.1.1. Show H-K $\int_0^3 f(x)\,dx = 3$, given

$$f(x) = \begin{cases} 4 & x = 0, \\ 1 & x \neq 0. \end{cases}$$

Hint:

$$\delta(x) = \begin{cases} \epsilon / \{2[|f(0)| + 1]\} & x = 0, \\ 1 & x \neq 0. \end{cases}$$

You will have noticed that both of these functions were integrated by ordinary Riemann techniques. We are trying to get used to the idea of a variable $\delta(\cdot)$. Let's try another example.

Example 8.1.2. Suppose

$$f(x) = \begin{cases} 1/x & x = 1, \frac{1}{2}, \frac{1}{3}, \dots, \\ 0 & \text{otherwise.} \end{cases}$$

Even though f is unbounded and thus not Riemann integrable, f is Lebesgue integrable ($f = 0$ a.e.), and area considerations suggest that HK $\int_0^1 f(x)\,dx = 0$.

Of course the only way a Riemann sum will be different from zero is if at least one tag c_k belongs to the set $\{1, \frac{1}{2}, \frac{1}{3}, \dots\}$; and such a tag could be the tag of two contiguous subintervals.

Let $\epsilon > 0$ be given. Then for any partition of $[a, b]$,

$$\left| \sum f(c_k)(x_k - x_{k-1}) - 0 \right| = \sum f(c_k)(x_k - x_{k-1}).$$

If $c_k \neq 1/n$ for all n, then this Riemann sum is zero. If $c_k = 1/j$, then

$$f(c_k)(x_k - x_{k-1}) = j(x_k - x_{k-1}) < j2\delta(c_k).$$

In that case, let's require

$$\delta(c_k) = \delta\left(\frac{1}{j}\right) = \frac{\epsilon}{j \cdot 2^{j+2}}.$$

Since a partition may assume only a finite number of values from the set $1, \frac{1}{2}, \frac{1}{3}, \ldots$, and a tag may be the tag of two subintervals, the difference between any associated Riemann sum and zero will be bounded by

$$\sum 4 \left(\frac{j\epsilon}{j \cdot 2^{j+2}} \right) = \epsilon.$$

It appears that

$$\delta(x) \equiv \begin{cases} \epsilon/n2^{n+2} & x = 1/n, \ n = 1, 2, \ldots, \\ 1 & \text{otherwise} \end{cases}$$

would work.

Exercise 8.1.2. Show that H-K $\int_0^1 f(x)\, dx = 0$, given that f is the Dirichlet function,

$$f(x) = \begin{cases} 1 & x \text{ rational}, \\ 0 & \text{otherwise}. \end{cases}$$

Hint:

$$\delta(x) = \begin{cases} \epsilon/2^{n+2} & x = r_n, \ n = 1, 2, \ldots, \\ 1 & \text{otherwise}. \end{cases}$$

Example 8.1.3. Consider the integral of a familiar function, $f(x) = x^2$, with $0 \leq x \leq 1$. Then

$$\text{C} \int_0^1 f(x)\, dx = \text{R} \int_0^1 f(x)\, dx = \text{L} \int_0^1 f\, d\mu = \frac{1}{3}.$$

Thus H-K $\int_0^1 f(x)\, dx$ should be $\frac{1}{3}$.

Now, f changes least about 0 and most about 1. Our function $\delta(\cdot)$ should reflect this behavior. We want to approximate the area under the curve between x_{k-1} and x_k by $c_k^2(x_k - x_{k-1})$. That is,

$$c_k^2(x_k - x_{k-1}) \approx \frac{1}{3} \left(x_k^3 - x_{k-1}^3 \right),$$

since we know the answer. Furthermore, the total error,

$$\sum_{k=1}^n \left[c_k^2 (x_k - x_{k-1}) - \frac{1}{3} \left(x_k^3 - x_{k-1}^3 \right) \right],$$

needs to be made small. But

$$x_{k-1}^2 (x_k - x_{k-1}) < \left(\frac{x_{k-1}^2 + x_k x_{k-1} + x_k^2}{3} \right) (x_k - x_{k-1})$$

$$= \frac{1}{3} \left(x_k^3 - x_{k-1}^3 \right)$$

$$< x_k^2 (x_k - x_{k-1}) .$$

Then $c_k^2(x_k - x_{k-1}) - \frac{1}{3} \left(x_k^3 - x_{k-1}^3 \right)$ is between

$$c_k^2(x_k - x_{k-1}) - x_k^2(x_k - x_{k-1}) \text{ and } c_k^2(x_k - x_{k-1}) - x_{k-1}^2(x_k - x_{k-1}).$$

That is, the individual errors are bounded by

$$(x_k - x_{k-1})(1 + c_k)(x_k - x_{k-1}) < 2\delta(c_k)(1 + c_k)(x_k - x_{k-1}),$$

and the cumulative error is bounded by

$$\sum_{k=1}^{n} 2\delta(c_k)(1 + c_k)(x_k - x_{k-1}),$$

which we want to be less than ϵ.

Would it work to set

$$\delta(x) = \epsilon/[2(1 + x)]?$$

Let's see:

$$\sum_{k-1}^{n} 2\delta(c_k)(1 + c_k)(x_k - x_{k-1}) < \sum_{k-1}^{n} 2\left(\frac{\epsilon}{2(1 + c_k)} \right)(1 + c_k)(x_k - x_{k-1})$$

$$= \epsilon.$$

Yes.

Comment: Consider

$$\delta(0) = \epsilon/2 > \epsilon/4 = \delta(1).$$

Notice that $\delta(\cdot)$ is a decreasing function. Thus the largest value of $\delta(\cdot)$, where the function changes least, is $x = 0$; the smallest value of $\delta(\cdot)$, where the function changes most, is $x = 1$. The subintervals of such a partition decrease in length as we move from left to right.

All of these *probablys* and *maybes* may be making us uneasy. In the next section we will tighten up our arguments and give precise definitions.

8.2   Gauges and δ-fine Partitions

To begin this discussion, we must establish some terminology.

- A positive function $\delta : [a, b] \to R^+$, where $\delta(t) > 0$ and $a \leq t \leq b$, is called a *gauge* on $[a, b]$.

- A *tagged partition* of $[a, b]$ is a finite collection of point intervals $(c_k, [x_{k-1}, x_k])$ where $1 \leq k \leq n$, $x_{k-1} \leq c_k \leq x_k$, and $a = x_0 < x_1 < x_2 < \cdots < x_n = b$.

- We call c_k the *tag of the interval* $[x_{k-1}, x_k]$.

- Given a gauge $\delta(\cdot)$ on $[a, b]$ and a tagged partition of $[a, b]$, we say the tagged partition is δ-*fine* if $c_k - \delta(c_k) < x_{k-1} \leq c_k \leq x_k < c_k + \delta(c_k)$, where $1 \leq k \leq n$.

To say we have a δ-fine partition of $[a, b]$ means we have a tagged partition of $[a, b]$ satisfying $c_k - \delta(c_k) < x_{k-1} \leq c_k \leq x_k < c_k + \delta(c_k)$ for each subinterval $[x_{k-1}, x_k]$.

Note that $x_k - x_{k-1} < 2\delta(c_k)$, where $1 \leq k \leq n$.

In the examples and exercises above, knowing the specific function (whether it was Dirichlet's or the "square" function) allowed us to construct a gauge. The question arises: Given an interval $[a, b]$ and a gauge $\delta(\cdot)$ on $[a, b]$, does there always exist a δ-fine partition of $[a, b]$? In other words, is there a partition so that $c_k - \delta(c_k) < x_{k-1} \leq c_k \leq x_k < c_k + \delta(c_k)$ for every subinterval in the partition?

Cousin's Lemma, below, addresses the existence of δ-fine partitions.

Lemma 8.2.1 (Cousin, 1895). *If δ is a gauge on $[a, b]$ (i.e., a positive function on $[a, b]$), then there exists a δ-fine partition of $[a, b]$. In fact, infinitely many exist.*

Cousin's argument uses nested intervals. Suppose we do not have a δ-fine partition of $[a, b]$. Then either $[a, (a + b)/2]$ or $[(a + b)/2, b]$ does not have a δ-fine partition, say $[a_1, b_1]$.

Repeat this bisction method. We have a sequence of nested intervals whose lengths approach zero. Consequently, we have a point c in the intersection; that is, $c \in [a_n, b_n]$ for all n. Because $\delta(c) > 0$, we have a natural number N so that $c - \delta(c) < a_N \leq c \leq b_N < c + \delta(c)$. But then we have a δ-fine partition of $[a_N, b_N]$ consisting of the tagged partition $(c, [a_N, b_N])$, a contradiction.

Note: If $0 < \delta(x) \leq \delta^*(x)$ on $[a, b]$, then any δ-fine partition of $[a, b]$ is a δ^*-fine partition of $[a, b]$. Why?

8.3   H-K Integrable Functions

We are now in a position to define the new integral.

Definition 8.3.1 (Henstock–Kurzweil Integral). A function f on the interval $[a, b]$ is said to be *Henstock–Kurzweil (H-K) integrable* on $[a, b]$ if there is a number A with the following property: For each $\epsilon > 0$ there exists a gauge (positive function) $\delta_\epsilon(\cdot)$ defined on $[a, b]$ such that for any δ_ϵ-fine partition of $[a, b]$, with $c_k - \delta_\epsilon(c_k) < x_{k-1} \leq c_k \leq x_k < c_k + \delta_\epsilon(c_k)$, for $1 \leq k \leq n$, we have

$$\left| \sum_{k=1}^{n} f(c_k)(x_k - x_{k-1}) - A \right| < \epsilon.$$

We write H-K $\int_a^b f(x)\, dx = A$.

As always, to say f is H-K integrable will mean that the integral exists and is finite. The sum $\sum f(c)\, \Delta x$ will be called an *H-K sum* when we are discussing the H-K integral.

Exercise 8.3.1. Show that a function f has at most one H-K integral. Assume that H-K $\int_a^b f(x)\, dx = A_i$, where $i = 1, 2$, and show that $A_1 = A_2$. Hint: $\delta(x) \equiv \min\{\delta_1(x), \delta_2(x)\}$, where δ_i is associated with A_i.

Just what kinds of functions are H-K integrable? We have discussed a few examples. A more thorough and systematic treatment will now be given.

8.3.1   Step Functions

Step functions are H-K integrable.

Example 8.3.1. Suppose the step function s is defined by

$$s(x) = \begin{cases} \alpha_1 & \text{if } a \leq x < c, \\ \beta & \text{if } x = c, \\ \alpha_2 & \text{if } c < x \leq b. \end{cases}$$

We will show that s is H-K integrable and

$$\text{H-K} \int_a^b s(x)\, dx = \alpha_1(c - a) + \alpha_2(b - c).$$

A possible difficulty arises at $x = c$, where s may be discontinuous. Let $\epsilon > 0$ be given. We will define a gauge $\delta_\epsilon(\cdot)$ on $[a, b]$ so that for any δ_ϵ-fine partition of $[a, b]$ the tag of the subinterval containing c — say, $[x_{k-1}, x_k]$ — is c. That is, $(c_k, [x_{k-1}, x_k]) = (c, [x_{k-1}, x_k])$. Define

$$\delta_\epsilon(x) = \begin{cases} \frac{1}{2}|x - c| & \text{if } x \neq c, \\ \delta & \text{if } x = c, \end{cases}$$

with δ to be determined. We claim $c = c_k$. Otherwise, $c_k \neq c$ and $|c - c_k| < 2\delta_\epsilon(c_k) < |c_k - c|$, a contradiction.

Now form the associated H-K sum $\big(s(c) = \beta\big)$ for a δ_ϵ-fine partition $[a, b]$:

$$\sum s(c_k)\, \Delta x = \alpha_1(x_{k-1} - a) + \beta(x_k - x_{k-1}) + \alpha_2(b - x_k)$$
$$= \alpha_1(c - a) + \alpha_2(b - c) + (\beta - \alpha_1)(c - x_{k-1})$$
$$+ (\beta - \alpha_2)(x_k - c).$$

Thus,

$$\left| \sum s(c_k)\Delta x - [\alpha_1(c - a) + \alpha_2(b - c)] \right| \leq |\beta - \alpha_1| \cdot 2\delta_\epsilon(c)$$

We want this last expression to be less than ϵ. Let

$$\delta_\epsilon(c) = \frac{\epsilon}{2(|\beta - \alpha_1| + |\beta - \alpha_2| + 1)} = \frac{\epsilon}{2(\text{sum of jumps} + 1)}.$$

That is, our gauge $\delta_\epsilon(\cdot)$ on $[a, b]$ is given by

$$\delta_\epsilon(x) = \begin{cases} \frac{1}{2}|x - c| & \text{if } x \neq c, \\ \epsilon/[2(\text{sum of jumps} + 1)] & \text{if } x = c. \end{cases}$$

We have shown that the step function is H-K integrable on $[a, b]$ and

$$\text{H-K} \int_a^b s(x)\, dx = \alpha_1(c - a) + \alpha_2(b - c).$$

Next, a more complicated step function.

Example 8.3.2. Let a step function s be defined by

$$s(x) = \begin{cases} \beta_0 & \text{if } x = a, \\ \alpha_1 & \text{if } a < x < c, \\ \beta_1 & \text{if } x = c, \\ \alpha_2 & \text{if } c < x \leq b, \\ \beta_2 & \text{if } x = b. \end{cases}$$

We will show that s is H-K integrable on $[a, b]$ and

$$\text{H-K} \int_a^b s(x)\, dx = \alpha_1(c - a) + \alpha_2(b - c).$$

Discontinuities of s may occur at a, c, or b. As in the previous example, the technique is to define a gauge that forces a, c, and b to be tags of their associated subintervals. Let $\epsilon > 0$ be given. Trial and error suggests an appropriate gauge on $[a, b]$ as

$$\delta_\epsilon(x) = \begin{cases} \frac{1}{2}\min\{|x-a|, |x-b|, |x-c|\} & \text{if } x \neq a, b, c, \\ \frac{1}{2}\min\{(c-a), (b-c), \delta\} & \text{if } x = a \text{ or } b \text{ or } c, \end{cases}$$

with δ to be determined.

We claim that a is the tag of $[a, x_1]$; b is the tag of $[x_{n-1}, b]$; and c is the tag of $[x_{k-1}, x_k]$ for some k, where $2 \leq k \leq n-1$, for any δ_ϵ-fine partition of $[a, b]$.

For example, to show a is the tag of $[a, x_1]$ when $a = c_1$, assume otherwise. Then $c_1 - \delta_\epsilon(c_1) < a$, so $c_1 - a < \delta_\epsilon(c_1)$. If c_1 is not b or c, then $c_1 - a < \delta_\epsilon(c_1) \leq \frac{1}{2}(c_1 - a)$, a contradiction. If $c_1 = b$, then $b - a < \delta_\epsilon(b) = \frac{1}{2}(b - c)$ and $a > c$. Finally, if $c_1 = c$, then $c - a < \delta_\epsilon(c) \leq \frac{1}{2}(c - a)$.

We have shown that $(c_1, [a, x_1]) = (a, [a, x_1])$. The reader may consider the case $(c_n, [x_{n-1}, b]) = (b, [x_{k-1}, b])$.

As for c, can it belong to the subinterval $[a, x_1]$? If so, then $c - a < 2\delta_\epsilon(a) \leq c - a$. Can c belong to the subinterval $[x_{n-1}, b]$? If so, then $b - c < 2\delta_\epsilon(b) \leq b - c$. Thus c belongs to $[x_{k-1}, x_k]$, where $2 \leq k \leq n-1$.

The tag c_k of $[x_{k-1}, x_k]$ is not a or b. If c_k is not c, then $|c - c_k| < 2\delta(c_k) < |c_k - c|$. The tag of $[x_{k-1}, x_k]$ must be c.

Form the associated H-K sum:

$$\sum s(c_k)\, \Delta x$$
$$= \alpha_1(c-a) + \alpha_2(b-c) + (\beta_0 - \alpha_1)(x_1 - a) + (\beta_1 - \alpha_1)(c - x_{k-1})$$
$$+ (\beta_1 - \alpha_2)(x_k - c) + (\beta_2 - \alpha_2)(b - x_{n-1}).$$

Thus

$$\left| \sum s(c_k)\, \Delta x - [\alpha_1(c-a) + \alpha_2(b-c)] \right|$$
$$\leq |\beta_0 - \alpha_1| \cdot 2\delta_\epsilon(a) + |\beta_1 - \alpha_1| \cdot 2\delta_\epsilon(c)$$
$$+ |\beta_1 - \alpha_2| \cdot 2\delta_\epsilon(c) + |\beta_2 - \alpha_2| \cdot 2\delta_\epsilon(b).$$

This last expression is less than ϵ if we choose

$$\delta = \frac{\epsilon}{2(\text{sum of jumps } + 1)}.$$

For the gauge $\delta_\epsilon(\cdot)$ on $[a, b]$ defined by

$$
\delta_\epsilon(x) = \begin{cases} \frac{1}{2} \min\{|x - a|, |x - b|, |x - c|\} & \text{if } x \neq a, b, c, \\ \frac{1}{2} \min\left\{(c - a), (b - c), \frac{\epsilon}{2(\text{sum of jumps} +1)}\right\} & \text{if } x = a, b \text{ or } c, \end{cases}
$$

the associated H-K sum satisfies

$$
\left| \sum s(c_k) \, \Delta x - [\alpha_1(c - a) + \alpha_2(b - c)] \right| < \epsilon.
$$

We have shown that this step function is H-K integrable on $[a, b]$ and

$$
\text{H-K} \int_a^b s(x) \, dx = \alpha_1(c - a) + \alpha_2(b - c).
$$

Extend these arguments to a general step function on $[a, b]$.

So, a finite number of discontinuities does not pose a problem. However, we know that step functions are Riemann and Lebesgue integrable. Furthermore,

$$
\text{R} \int_a^b s(x) \, dx = \text{L} \int_a^b s \, d\mu = \text{H-K} \int_a^b s(x) \, dx.
$$

Have we anything new? This question will be answered in Sections 8.7.1 and 8.7.3. But first we have some preliminary work to do.

8.3.2   Riemann Integrable Functions

Riemann integrable functions are H-K integrable, and the integrals have the same value.

Suppose f is Riemann integrable on $[a, b]$. Let $\epsilon > 0$ be given. We have a positive constant δ so that if P is any partition of $[a, b]$ with $x_k - x_{k-1} < \delta$, then

$$
\left| \sum_P f(c_k)(x_k - x_{k-1}) - \text{R} \int_a^b f(x) \, dx \right| < \epsilon.
$$

Define our gauge δ_ϵ by $\delta_\epsilon(x) = \delta/2$ for all x in $[a, b]$. Suppose we take any δ_ϵ-fine partition of $[a, b]$, the existence of which is guaranteed by Cousin's Lemma 8.2.1. Then $z_j - z_{j-1} < 2\delta(c_j) < \delta$, and consequently

$$
\left| \sum f(c_j)(z_j - z_{j-1}) - \text{R} \int_a^b f(x) \, dx \right| < \epsilon.
$$

The function f is H-K integrable and H-K $\int_a^b f(x)\,dx = $ R $\int_a^b f(x)\,dx$.

Of course this result makes the step function example of Section 8.3.1 trivial.

Example 8.3.3. The function

$$f(x) = \begin{cases} 1/\sqrt{x} & 0 < x \leq 1, \\ 0 & x = 0, \end{cases}$$

is H-K integrable on $[0, 1]$ and H-K $\int_0^1 f(x)\,dx = 2$.

How shall we demonstrate this? We begin by looking at the fluctuations. Because of the rapid change in f about zero, we want a small interval about zero. The area under the curve between x_{k-1} and x_k is given by $2\sqrt{x_k} - 2\sqrt{x_{k-1}}$. Thus we want

$$\frac{1}{\sqrt{c_k}}(x_k - x_{k-1}) \approx 2\sqrt{x_k} - 2\sqrt{x_{k-1}} \qquad \text{for } c_k > 0.$$

If $c_1 = 0$, then $0(x_1 - 0) \approx 2\sqrt{x_1}$ and $2\sqrt{x_1} < 2\sqrt{\delta(x)} \leq \epsilon/2$, where we assume $\epsilon < 4$. So, $\delta(0) = \epsilon^2/16$.

Otherwise, for $k = 2, \ldots, n$, routine manipulation shows

$$\left| \frac{1}{\sqrt{c_k}}(x_k - x_{k-1}) - 2\left(\sqrt{x_k} - \sqrt{x_{k-1}}\right) \right|$$

$$= \left| \frac{1}{\sqrt{c_k}}(x_k - c_k + c_k - x_{k-1}) - 2\left(\sqrt{x_k} - \sqrt{c_k} + \sqrt{c_k} - \sqrt{x_{k-1}}\right) \right|$$

$$\leq \left| \frac{1}{\sqrt{c_k}}(x_k - c_k) - 2\left(\sqrt{x_k} - \sqrt{c_k}\right) \right|$$

$$\quad + \left| \frac{1}{\sqrt{c_k}}(c_k - x_{k-1}) - 2\left(\sqrt{c_k} - \sqrt{x_{k-1}}\right) \right|$$

$$= (x_k - c_k) \left| \frac{\sqrt{x_k} - \sqrt{c_k}}{\sqrt{c_k}\left(\sqrt{x_k} + \sqrt{c_k}\right)} \right| + (c_k - x_{k-1}) \left| \frac{\sqrt{c_k} - \sqrt{x_{k-1}}}{\sqrt{c_k}\left(\sqrt{c_k} + \sqrt{x_{k-1}}\right)} \right|$$

$$\leq (x_k - c_k) \frac{\sqrt{x_k} - \sqrt{c_k}}{c_k} + (c_k - x_{k-1}) \frac{\sqrt{c_k} - \sqrt{x_{k-1}}}{c_k}$$

$$\leq (x_k - c_k) \frac{x_k - c_k}{c_k^{3/2}} + (c_k - x_{k-1}) \frac{c_k - x_{k-1}}{c_k^{3/2}}$$

$$\leq \frac{2(x_k - x_{k-1})^2}{c_k^{3/2}} < \frac{2\delta(c_k)}{c_k^{3/2}}(x_k - x_{k-1}).$$

Thus, the cumulative error for the interval $[x_1, 1]$ is bounded by

$$\sum_{k=2}^{n} \frac{2\delta(c_k)}{c_k^{3/2}}(x_k - x_{k-1}),$$

which we want to be less than $\epsilon/2$. We have this if $2\delta(c_k)/c_k^{3/2} < \epsilon/2$; that is, if $\delta(c_k) < \epsilon/4c_k^{3/2}$ for $k = 2, \ldots, n$. This suggests

$$\delta(x) = \begin{cases} \epsilon^2/16 & x = 0, \\ (\epsilon/4)x^{3/2} & 0 < x \le 1. \end{cases}$$

Will this gauge work? Let $\epsilon > 0$ be given (we may assume $\epsilon < 4$), with

$$\delta_\epsilon(x) \equiv \begin{cases} \epsilon^2/16 & x = 0, \\ (\epsilon/4)x^{3/2} & 0 < x \le 1. \end{cases}$$

Let P be any δ_ϵ-fine partition of $[0, 1]$. (Once again, its existence is guaranteed by Cousin's Lemma 8.2.1.)

Consider the collection of associated point intervals: $(c_1, [0, x_1])$, $(c_2, [x_1, x_2])$, $\ldots$, $(c_n, [x_{n-1}, x_n])$. We claim that for the first interval, $[0, x_1]$, the tag c_1 must be zero.

If c_1 is greater than zero, then $\delta_\epsilon(c_1) = (\epsilon/4)c_1^{3/2}$, and $c_1 - (\epsilon/4)c_1^{3/2}$ must be less than zero. But

$$c_1 - \frac{\epsilon}{4}c_1^{3/2} = \frac{c_1\left(4 - \epsilon c_1^{1/2}\right)}{4} \ge \frac{c_1(4 - \epsilon)}{4} > 0.$$

So the tag of the first subinterval is zero. Thus,

$$\left| \sum_{k=1}^{n} f(c_k)(x_k - x_{k-1}) - 2 \right|$$

$$= \left| \sum_{k=2}^{n} f(c_k)(x_k - x_{k-1}) - 2 \right|$$

$$= \left| \sum_{k=2}^{n} \frac{1}{\sqrt{c_k}}(x_k - x_{k-1}) - 2\left(\sqrt{x_k} - \sqrt{x_{k-1}}\right) - 2\left(\sqrt{x_1} - \sqrt{0}\right) \right|$$

$$\le 2\sqrt{x_1} + \sum_{k=2}^{n} \left| \frac{1}{\sqrt{c_k}}(x_k - x_{k-1}) - 2\left(\sqrt{x_k} - \sqrt{x_{k-1}}\right) \right|$$

$$\le 2\sqrt{\delta(0)} + \sum_{k=2}^{n} 2\frac{\delta(c_k)}{c_k^{3/2}}(x_k - x_{k-1})$$

$$\le 2\left(\frac{\epsilon}{4}\right) + \sum_{k=2}^{n} 2\frac{\epsilon}{4}\frac{c_k^{3/2}}{c_k^{3/2}}(x_k - x_{k-1}) < \frac{\epsilon}{2} + \frac{\epsilon}{2} = \epsilon.$$

So f is H-K integrable on $[0, 1]$ and H-K $\int_0^1 f(x)\,dx = 2$. This function is not Riemann integrable, although it is Lebesgue integrable.

Example 8.3.4. A function that is unbounded on every subinterval of $[0, 1]$, with

$$\text{H-K} \int_0^1 f(x)\,dx = \text{L} \int_0^1 f\,d\mu = 0.$$

As in Exercise 6.2.3, let

$$f(x) = \begin{cases} q & x = p/q; \, p, q \text{ relatively prime natural numbers,} \\ 0 & \text{otherwise.} \end{cases}$$

Recall Exercise 8.1.2 and Example 8.1.2.

Example 8.3.5. Suppose that f is zero almost everywhere on $[a, b]$. Then f is H-K integrable on $[a, b]$ and

$$\text{H-K} \int_a^b f(x)\,dx = \text{L} \int_a^b f\,d\mu = 0.$$

Let $\epsilon > 0$ and $E = \{x \in [a, b] \mid f(x) \neq 0\}$. Because E has measure zero, we can cover E by an open set G so that $\mu(G) < \epsilon$.

To start, suppose we have a Riemann partition of $[a, b]$. The only contribution to this Riemann sum will occur when a tag c_k belongs to $E \subset G$. How do we estimate $f(c_k)$? All we know is that $f(c_k) \neq 0$.

The idea is to partition E into disjoint measurable sets, any one of which would yield an estimate for f. One way to do this is to observe that

$$E = \{x \in [a, b] \mid f(x) \neq 0\} = \{x \in [a, b] \mid |f(x)| \neq 0\}$$
$$= \cup\{x \in [a, b] \mid n - 1 < |f(x)| \leq n\}.$$

Each of the mutually disjoint sets E_n, $\{x \in [a, b] \mid n - 1 < |f(x)| \leq n\}$, is a subset of E, which has measure zero. The set E_n is measurable, and $\mu(E_n) = 0$.

We have an open set G_n covering E_n with $\mu(G_n) < \epsilon/n2^n$ (recall Section 5.6 on approximating measurable sets). Now we need a gauge.

If a tag c_k belongs to $[a, b] - E$, then $f(c_k) = 0$ and we would have no contribution to a Riemann sum. In this case, $\delta_\epsilon(c_k)$ does not matter. Otherwise, c_k belongs to E and we have a unique N so that $c_k \in E_N$.

Since $|f(c_k)| > N - 1$, which is potentially very large, we need the subintervals associated with such c_k, $[x_{k-1}, x_k]$, to be small. Recalling that

E_N is a subset of the open set G_N, and considering that $\mu(G_N)$ is less than $\epsilon/N2^N$, is there some way to force the nonoverlapping subintervals $[x_{k-1}, x_k]$ to be a subset of G_N and thus to have a total length not exceeding $\epsilon/N2^N$?

Since $[x_{k-1}, x_k] \subset (c_k - \delta(c_k), c_k + \delta(c_k))$, is there some way to force $(c_k - \delta(c_k), c_k + \delta(c_k))$ to be in G_N? Yes. Let $\delta(c_k)$ be the distance from c_k to points not in G_N. That is, let $\delta(c_k)$ equal the distance from c_k to points in the complement of the open set G_N.

We are ready to define the gauge:

$$\delta_\epsilon(x) = \begin{cases} 1 & x \in [a, b] - E, \\ \text{distance from } x \text{ to the complement of } G_n & \begin{array}{l} \text{for } x \in E_n, \\ n = 1, 2, \ldots. \end{array} \end{cases}$$

Consider a δ_ϵ-fine partition of $[a, b]$. Only tags in E_n, for some n, will contribute: $E_{n_1}, \ldots, E_{n_m}$. Because $|f| \leq n_i$ on E_{n_i}, the total contribution is bounded by

$$n_1 \mu(G_{n_1}) + n_2 \mu(G_{n_2}) + \cdots + n_k \mu(G_{n_m}) < \sum n\mu(G_n) < \epsilon.$$

Thus, H-K $\int_a^b f(x)\, dx = 0$. Look at Example 8.3.4.

We offer two additional observations. First, if C is the Cantor function, then $C' = 0$ almost everywhere and H-K $\int_0^1 C'(x)\, dx = 0 = \text{L} \int_0^1 C'\, d\mu$.

Second, almost everywhere equal functions have the same integral.

Exercise 8.3.2. Suppose f is H-K integrable on $[a, b]$ and g equals f almost everywhere on $[a, b]$. Demonstrate that g is H-K integrable on $[a, b]$ and that

$$\text{H-K} \int_a^b g(x)\, dx = \text{H-K} \int_a^b f(x)\, dx.$$

Hint: $g = (g - f) + f$; $g - f$ is H-K integrable by Example 8.3.5; and by the same example we have H-K $\int_a^b [g(x) - f(x)]\, dx = 0$. Now use linearity of the integral.

Is the H-K integral an extension of the Lebesgue integral? We have some preliminary work to do before we can answer this question.

8.4   The Cauchy Criterion for H-K Integrability

As with other integration processes, Cauchy-type conditions are useful. An H-K sum is a Riemann sum associated with a δ_ϵ-fine partition.

Definition 8.4.1 (Cauchy Criterion for H-K Integrability). A function f on the interval $[a, b]$ is H-K integrable on $[a, b]$ iff for every $\epsilon > 0$ we have δ_ϵ-fine partitions P_1 and P_2 of $[a, b]$ with the associated H-K sums within ϵ of each other.

Proof. Let $\epsilon > 0$ be given. If f is H-K integrable on $[a, b]$, then we have a gauge $\delta_\epsilon(\cdot)$ on $[a, b]$ so that the H-K sum of any δ_ϵ-fine partition of $[a, b]$ is within $\epsilon/2$ of the H-K integral of f. Thus any two such sums will be within ϵ of each other.

For the other direction, assume that for any $\epsilon > 0$ we can determine a gauge $\delta_\epsilon(\cdot)$ so that for any two δ-fine partitions of $[a, b]$ the H-K sums are within ϵ of each other. Then for $\epsilon = 1/n$, we have a positive function $\delta_n(\cdot)$ so that for any two δ_n-fine partitions the associated H-K sums are within $1/n$ of each other and $\delta_{n+1} < \delta_n$ on $[a, b]$.

Consider the sequence of the H-K sums

$$\sum_{P_1} f(c^1)\Delta_1 x, \sum_{P_2} f(c^2)\Delta_2 x, \ldots, \sum_{P_n} f(c^n)\Delta_n x, \ldots,$$

with P_n a δ_n-fine partition of $[a, b]$. We claim the sequence

$$\left\{ \sum_{P_n} f(c^n)\Delta_n x \right\}$$

is a Cauchy sequence of real numbers. To show this, we compare

$$\sum_{P_n} f(c^n)\Delta_n x \quad \text{and} \quad \sum_{P_m} f(c^m)\Delta_m x$$

with $m > n$. The key is the observation that by our construction process $\delta_m < \delta_n$, so a δ_m-fine partition is a δ_n-fine partition of $[a, b]$. We have two H-K sums for a δ_n-fine partition of $[a, b]$. For $m > n$,

$$\left| \sum_{P_m} f(c^m)\Delta_m x - \sum_{P_n} f(c^n)\Delta_n x \right| < \frac{1}{n}$$

and we have a Cauchy sequence. Let $A = \lim \sum_{P_n} f(c^n)\Delta_n x$. Then for each n

$$\left| \sum_{P_n} f(c^n)\Delta_n x - A \right| \leq \frac{1}{n}.$$

Let $\epsilon > 0$ be given and choose N so that $1/N < \epsilon$. Let P be any δ_N-fine partition of $[a, b]$. Then

$$\left| \sum_P f(c)\, \Delta x - A \right| \leq \left| \sum_P f(c)\, \Delta x - \sum_{P_N} f(c^N)\Delta_N x \right|$$

$$+ \left| \sum_{P_N} f(c^N)\Delta_N x - A \right|$$

$$< \frac{1}{N} + \frac{1}{N} < 2\epsilon,$$

since P and P_N are δ_N-fine partitions of $[a, b]$.

Thus f is H-K integrable on $[a, b]$ and H-K $\int_a^b f(x)\, dx = A$. $\quad\square$

Another way to look at H-K integrability involves approximation above and below. We will explore this by way of an example.

Example 8.4.1. Suppose for each $\epsilon > 0$ we have H-K integrable functions ϕ and ψ on $[a, b]$ so that $\phi \leq f \leq \psi$ and

$$\text{H-K} \int_a^b \psi(x)\, dx - \text{H-K} \int_a^b \phi(x) < \epsilon.$$

Then f is H-K integrable on $[a, b]$.

Before we demonstrate this, notice that since ϕ and ψ are H-K integrable, we have δ_ϕ- and δ_ψ-fine partitions of $[a, b]$, $\phi \leq f \leq \psi$ on $[a, b]$ so that

$$\left| \sum_{P_\phi} \phi(c)\, \Delta x - \text{H-K} \int_a^b \phi(x)\, dx \right|$$

and

$$\left| \sum_{P_\psi} \psi(d)\Delta y - \text{H-K} \int_a^b \psi(x)\, dx \right|$$

are less than ϵ.

Define a gauge $\delta(\cdot)$ by $\delta(x) = \min\{\delta_\phi(x), \delta_\psi(x)\}$ and let P be any δ-fine partition of $[a, b]$. Then a δ-fine partition of $[a, b]$ will be a δ_ϕ- and

δ_ψ-fine partition of $[a, b]$, and

$$\text{H-K} \int_a^b \phi(x)\, dx - \epsilon < \sum_P \phi(c)\, \Delta x$$

$$\leq \sum_P f(c)\, \Delta x \leq \sum_P \psi(c)\, \Delta x$$

$$< \text{H-K} \int_a^b \psi(x)\, dx + \epsilon$$

$$< \text{H-K} \int_a^b \phi(x)\, dx + 2\epsilon.$$

Any two H-K sums for a δ-fine partition of $[a, b]$ will be within 3ϵ of each other. By the Cauchy Criterion (Definition 8.4.1), f is H-K integrable on $[a, b]$.

We conclude this discussion with a final property.

Example 8.4.2. If f is H-K integrable on $[a, b]$, then f is H-K integrable on every closed subinterval $[c, d]$ of $[a, b]$.

Suppose $a < c < d < b$. (The reader may consider the other cases.) The idea here is to construct two δ_ϵ-fine partitions of $[a, b]$ that are identical on $[a, c]$ and $[d, b]$, and then use Cauchy's Criterion.

Let $\epsilon > 0$ be given. Because f is H-K integrable on $[a, b]$, we have a gauge δ_ϵ on $[a, b]$. Thus we have a gauge δ_ϵ on $[a, c]$, $[c, d]$, and $[d, b]$. By Cousin's Lemma 8.2.1), we have δ_ϵ-fine partitions of $[a, c]$, $[c, d]$, and $[d, b]$.

Let P_{ac}, P_{cd}^1, P_{cd}^2, P_{db} denote such δ_ϵ-fine partitions with P_{cd}^1, P_{cd}^2 any two δ_ϵ-fine partitions of $[c, d]$. We have

$$P_{ac}^1 = \{(c_1, [a, x_1]), \ldots, (c_N, [x_{N-1}, c])\},$$
$$P_{cd}^1 = \{(d_1, [c, z_1^1]), \ldots, (d_M, [z_{M-1}^1, d])\},$$
$$P_{cd}^2 = \{(e_1, [c, z_1^2]), \ldots, (e_L, [z_{L-1}^2, d])\},$$
$$P_{db} = \{(f_1, [d, y_1]), \ldots, (f_K, [y_{K-1}, c])\}.$$

Notice that $P_{ac} \cup P_{cd}^1 \cup P_{db}$ and $P_{ac} \cup P_{cd}^2 \cup P_{db}$ are two δ_ϵ-fine partitions of $[a, b]$.

By the Cauchy Criterion, the difference of the associated H-K sums is less than ϵ. But these H-K sums are identical on $[a, c]$ and $[d, b]$. Thus the difference of these H-K sums is a difference of any two δ_ϵ-fine partitions on $[c, d]$ that is less than ϵ. Using the Cauchy Criterion again we conclude that f is H-K integrable on $[c, d]$.

8.5 Henstock's Lemma

Another useful result, due to Ralph Henstock, tells us that good approximations over the entire interval yield good approximations over unions of subintervals.

Lemma 8.5.1 (Henstock, 1961). *Suppose f is H-K integrable on $[a, b]$, and for $\epsilon > 0$ let δ_ϵ be a gauge on $[a, b]$ so that if we have a δ_ϵ-fine partition of $[a, b]$, then*

$$\left| \sum f(c_k)\, \Delta x - \text{H-K} \int_a^b f(x)\, dx \right| < \epsilon.$$

Suppose $F_1, F_2, \ldots, F_J$ is a finite collection of nonoverlapping (no common interior) closed subintervals of $[a, b]$, with $y_j \in F_j \subset \left(y_j - \delta_\epsilon(y_j), y_j + \delta_\epsilon(y_j)\right)$, where $1 \leq j \leq J$. Then

$$\left| \sum_{j=1}^J \left\{ f(y_j)\ell(F_j) - \text{H-K} \int_{F_j} f(x)\, dx \right\} \right| \leq \epsilon$$

and

$$\sum_{j=1}^J \left| f(y_j)\ell(F_j) - \text{H-K} \int_{F_j} f(x)\, dx \right| \leq 2\epsilon,$$

where $\ell(F_j)$ denotes the length of the subinterval F_j.

Proof. The set $[a, b] - \cup_1^J F_j$ is a finite collection of open intervals. Adjoin their endpoints. For this finite collection of closed subintervals, $K_1, K_2, \ldots,$ K_N, of $[a, b]$, f is H-K integrable (Example 8.4.2).

Choose a gauge $\delta_n < \delta_\epsilon$ so that for $\eta > 0$ and a δ_n-fine partition of K_n, call it $K(P_n)$, we have

$$\left| \sum_{K(P_n)} f(x)\, \Delta x - \text{H-K} \int_{K_n} f(x)\, dx \right| < \frac{\eta}{N}.$$

The partitions $K(P_1), K(P_2), \ldots K(P_N)$, together with $F_1, F_2, \ldots, F_J$ form a δ_ϵ-fine partition of $[a, b]$. (See previous result.) We have the case as illustrated in Figure 1.

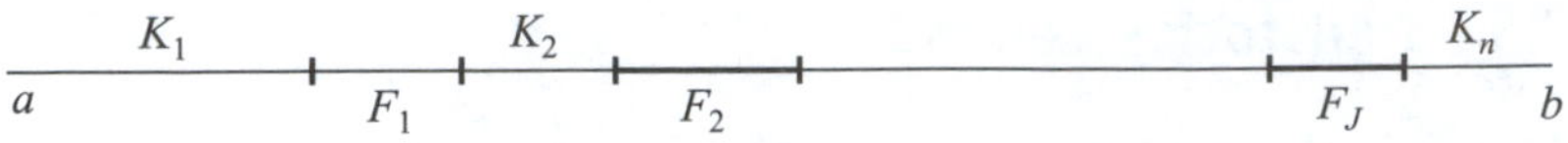

Figure 1. Partition of $[a, b]$

Then

$$\left| \sum_{j=1}^{J} \left\{ f(y_j)\ell(F_j) - \text{H-K} \int_{F_j} f(x)\, dx \right\} \right|$$

$$= \left| \left[f(y_1)\ell(F_1) + f(y_2)\ell(F_2) + \cdots + f(y_J)\ell(F_J) \right. \right.$$

$$\left. + \sum_{K(P_1)} f(x)\, \Delta x + \sum_{K(P_2)} f(x)\, \Delta x + \cdots + \sum_{K(P_N)} f(x)\, \Delta x \right]$$

$$- \text{H-K} \left[\int_{F_1} f(x)\, dx + \int_{F_2} f(x)\, dx + \cdots + \int_{F_J} f(x)\, dx \right.$$

$$\left. + \int_{K_1} f(x)\, dx + \int_{K_2} f(x)\, dx + \cdots + \int_{K_N} f(x)\, dx \right]$$

$$\left. + \sum_{n=1}^{N} \left(\text{H-K} \int_{K_n} f(x)\, dx \right) - \sum_{n=1}^{N} \left(\sum_{K(P_n)} f(x)\, \Delta x \right) \right|$$

$$< \epsilon + \sum_{n=1}^{N} \left| \text{H-K} \int_{K_n} f(x)\, dx - \sum_{K(P_n)} f(x)\, dx \right| < \epsilon + N \frac{\eta}{N} = \epsilon + \eta.$$

By the arbitrary nature of η, the first inequality is valid. For the second equality, select from the subintervals $F_1, F_2, \ldots, F_J$ those for which $f(y_i)\ell(F_j) - \text{H-K} \int_{F_j} f(x)\, dx \geq 0$. Thus

$$0 \leq \sum f(y_j)\ell(F_j) - \text{H-K} \int_{F_j} f(x)\, dx = \sum |\ \ | \leq \epsilon.$$

Otherwise, $f(y_j)\ell(F_j) - \text{H-K} \int_{F_J} f(x)\, dx < 0$, and

$$- \sum f(y_j)\ell(F_j) + \text{H-K} \int_{F_j} f(x)\, dx = \sum |\ \ | \leq \epsilon.$$

All together,

$$\sum \left| f(y_j)\ell(F_j) - \text{H-K} \int_{F_j} f(x)\, dx \right| \leq 2\epsilon.$$

The argument is complete. $\square$

8.6   Convergence Theorems for the H-K Integral

We now establish a convergence theorem for H-K integrals.

8.6.1   Monotone Convergence

Theorem 8.6.1 (H-K Monotone Convergence). *Suppose that* $\{f_k\}$ *is a monotone sequence of H-K integrable functions on* $[a, b]$ *converging pointwise to* f *on* $[a, b]$. *Then* f *is H-K integrable on* $[a, b]$ *iff the sequence* $\{$H-K $\int_a^b f_k(x)\, dx\}$ *is bounded on* $[a, b]$. *In this case,*

$$\text{H-K} \int_a^b f(x)\, dx = \lim \text{H-K} \int_a^b f_k(x)\, dx.$$

Proof. Assume the sequence $\{f_k\}$ is monotone increasing: $f_1 \leq \cdots \leq f_k \leq f_{k+1} \leq f$ on $[a, b]$. If f is H-K integrable, then H-K $\int_a^b f_k(x)\, dx \leq$ H-K $\int_a^b f(x)\, dx$ for all k, and the monotone increasing sequence $\{$H-K $\int_a^b f_k(x)\, dx\}$, being bounded above by H-K $\int_a^b f(x)\, dx$, converges.

Now, assume the monotone increasing sequence $\{$H-K $\int_a^b f_k(x)\, dx\}$ is bounded above and converges to A: $\lim$ H-K $\int_a^b f_k(x)\, dx = A$. We will show that H-K $\int_a^b f(x)\, dx = A$; that is, f is H-K integrable on $[a, b]$ to A.

Let $\epsilon > 0$ be given. For $k \geq K$, we have $0 \leq A -$ H-K $\int_a^b f_k(x)\, dx < \epsilon$. By the assumption of H-K integrability of f_k, we have a gauge δ_ϵ^k for f_k so that

$$\left| \sum f(c_k)\, \Delta x - \text{H-K} \int_a^b f_k(x)\, dx \right| < \frac{\epsilon}{2^k}$$

for every δ_ϵ^k-fine partition of $[a, b]$.

Let x be any point in $[a, b]$. Because $\lim f_k(x) = f(x)$, we have a natural number $n(x) \geq K$ so that $|f(x) - f_k(x)| < \epsilon$ whenever $k \geq n(x) \geq K$, a natural number associated with each point of $[a, b]$.

The function $\delta_\epsilon(x) = \delta_\epsilon^{n(x)}$ is a gauge on $[a, b]$. Suppose P is a δ_ϵ-fine partition of $[a, b]$. We will show that the difference of the associated H-K sum and the number A can be made arbitrarily small.

With $\sum_P f(c)\,\Delta x = \sum_{i=1}^{I} f(c_i)\,\Delta x_i$, we have

$$\left| \sum_P f(c)\,\Delta x - A \right| \le \left| \sum_{i=1}^{I} f(c_i)\,\Delta x_i - \sum_{i=1}^{I} f_{n(c_i)}(c_i)\,\Delta x_i \right|$$

$$+ \left| \sum_{i=1}^{I} \left\{ f_{n(c_i)}(c_i)\,\Delta x_i - \text{H-K} \int_{\Delta x_i} f_{n(c_i)}(x)\,dx \right\} \right|$$

$$+ \left| \sum_{i=1}^{I} \text{H-K} \int_{\Delta x_i} f_{n(c_i)}(x)\,dx - A \right|.$$

The first term on the right is dominated by $\sum_{i=1}^{I} \left| f(c_i) - f_{n(c_i)}(c_i) \right| \Delta x_i$, which is less than $\epsilon(b-a)$.

For the third term, recall that the sequence $\{f_k\}$ is monotone increasing. The natural numbers $n(c_1), n(c_2), \dots, n(c_I)$ are each greater than or equal to K. Then

$$\text{H-K} \int_a^b f_K(x)\,dx = \sum_{i=1}^{I} \int_{\Delta x_i} f_K(x)\,dx \le \sum_{i=1}^{I} \int_{\Delta x_i} f_{n(c_i)}(x)\,dx,$$

so

$$0 \le A - \sum_{i=1}^{I} \text{H-K} \int_{\Delta x_i} f_{n(c_i)}(x)\,dx \le A - \text{H-K} \int_a^b f_K(x)\,dx < \epsilon.$$

The second term remains:

$$\left| \sum_{i=1}^{I} \left\{ f_{n(c_i)}(c_i)\,\Delta x_i - \text{H-K} \int_{\Delta x_i} f_{n(c_i)}(x)\,dx \right\} \right|.$$

The natural numbers $n(c_1), n(c_2), \dots, n(c_I)$ may not be distinct. Those that are the same correspond to the same partition, and we use Henstock's Lemma. For example, if $n(c_{i_1}) = n(c_{i_2}) = \cdots = n(c_{i_l})$, then

$$\left| \sum_{k=1}^{\ell} f_{n(c_{i_k})}\,\Delta x_{i_k} - \text{H-K} \int_{\Delta x_{i_k}} f_{n(c_{i_k})}(x)\,dx \right| \le \frac{\epsilon}{2^{n(c_{i_l})}}.$$

The second term is dominated by $\sum \epsilon/2^k = \epsilon$. We have shown that for this δ_ϵ-fine partition of $[a,b]$ the difference between the associated H-K sum for the function f and the number A is bounded by $\epsilon(b-a) + \epsilon + \epsilon$.

This is what it means to say that f is H-K integrable on $[a, b]$ and that the H-K integral of f equals A:

$$\text{H-K} \int_a^b f(x)\, dx = \lim \text{H-K} \int_a^b f_k(x)\, dx.$$

The argument for $f_1 \geq \cdots \geq f_k \geq f_{k+1} \geq \cdots \geq f$ is similar. □

We note that pointwise convergence of f_k to f may be replaced by f_k convergence pointwise to f almost everywhere on $[a, b]$. We define f to be zero whenever pointwise convergence fails.

8.6.2   Dominated Convergence

We have the Henstock–Kurzweil counterpart to the Lebesgue Dominated Convergence Theorem 6.3.3.

Theorem 8.6.2 (H-K Dominated Convergence). *Suppose $\{f_k\}$ is a sequence of H-K integrable functions on $[a, b]$ converging pointwise to f on $[a, b]$. If we have H-K integrable functions ϕ and ψ such that $\phi \leq f_k \leq \psi$ for all k, then f is H-K integrable and*

$$\text{H-K} \int_a^b f(x)\, dx = \lim \text{H-K} \int_a^b f_k(x)\, dx.$$

The proof may be found in Gordon (1994) or Bartle (2001).

8.7   Some Properties of the H-K Integral

It is almost time to discuss Fundamental Theorems of Calculus for the H-K integral. Before doing so, however, we will explore some more of its properties.

8.7.1   Extension of Lebesgue

We begin by showing that Lebesgue integrable functions are H-K integrable. In fact, the H-K integral is an extension of the Lebesgue integral. Liberal use of the convergence theorems facilitates the argument.

Theorem 8.7.1. *Suppose f is Lebesgue integrable on $[a, b]$. Then f is H-K integrable on $[a, b]$ and*

$$\text{H-K} \int_a^b f(x)\, dx = \text{L} \int_a^b f\, d\mu.$$

Proof. Our argument has three parts. We begin by recalling that Section 8.3.1 showed that step functions are Lebesgue and H-K integrable and the integrals are equal.

Part 1. Simple functions are H-K integrable, and the H-K and Lebesgue integrals have the same value.

Let G be an open subset of $[a, b]$. It can be written as a countable union of disjoint intervals: $G = \cup I_k$. Define a sequence of step functions $\{f_k\}$ by $f_k = \chi_{I_1} + \chi_{I_2} + \cdots + \chi_{I_k}$. Then f_k is H-K integrable and $\lim f_k = \chi_G$.

Note that if $x \in [a, b] - G$, then $f_k(x) = 0$ for all k and $\lim f_k(x) = 0$. If $x \in G$, then we have a natural number N so that $x \notin I_1 \cup \cdots \cup I_{N-1}$, $x \in I_N$, $x \notin I_{N+1}, \ldots$. That is,

$$f_1(x) = \cdots = f_{N-1}(x) = 0,$$
$$f_N(x) = 1 = f_{N+1}(x) = \cdots, \quad \text{and}$$
$$\lim f_k(x) = 1.$$

By the Monotone Convergence Theorems (H-K and Lebesgue), since $0 \le f_k \le 1$ on $[a, b]$ the characteristic function on G is H-K and Lebesgue integrable, and

$$\text{H-K} \int_a^b \chi_G(x)\, dx = \lim \text{H-K} \int_a^b f_k(x)\, dx$$
$$= \lim \text{L} \int_a^b f_k\, d\mu = \text{L} \int_a^b \chi_G\, d\mu.$$

Next, let E be a measurable subset of $[a, b]$. We may cover E with a monotone decreasing sequence of open sets $\{G_k\}$ such that $\lim \mu(G_k) = \mu(E)$ (by Theorem 5.6.1). Again, $f_k = \chi_{G_k}$ defines a monotone sequence of H-K and Lebesgue integrable functions converging pointwise to χ_E, with $0 \le f_k \le 1$ on $[a, b]$ and $0 \le \text{H-K} \int_a^b f_k(x)\, dx \le b - a$ for all k. We have

$$\text{H-K} \int_a^b \chi_E(x)\, d\mu = \lim \text{H-K} \int_a^b f_k(x)\, dx$$
$$= \lim \text{L} \int_a^b f_k\, d\mu = \text{L} \int_a^b \chi_E\, d\mu.$$

Finally, a simple function is by definition a linear combination of characteristic functions on measurable sets. Thus,

$$\text{H-K} \int_a^b (\text{simple function})\, dx = \text{L} \int_a^b (\text{simple function})\, d\mu.$$

Part 2. Assume f is a nonnegative Lebesgue integrable function on $[a, b]$. Then we have a monotone sequence $\{\phi_k\}$ of simple functions converging pointwise to f on $[a, b]$ (Theorem 5.7.2). Now,

$$\text{H-K} \int_a^b \phi_k(x)\,dx = \text{L} \int_a^b \phi_k\,d\mu \leq \text{L} \int_a^b f\,d\mu < \infty.$$

Application of the Monotone Convergence Theorems yields

$$\text{H-K} \int_a^b f(x)\,dx = \lim \text{H-K} \int_a^b \phi_k(x)\,dx$$

$$= \lim \text{L} \int_a^b \phi_k\,d\mu = \text{L} \int_a^b f\,d\mu.$$

Part 3. Apply Part 2 to the nonnegative Lebesgue integrable functions

$$\frac{|f| + f}{2} \qquad \text{and} \qquad \frac{|f| - f}{2}. \qquad \square$$

So Lebesgue integrable functions are Henstock–Kurzweil integrable and the integrals have the same value. Are there functions that are H-K integrable and not Lebesgue integrable? We will show very shortly that there are.

8.7.2 Recovering Functions via Differentiation

We now prove a theorem that tells us that, for differentiable functions, the H-K integral "recovers" the function from its derivative. Recall the Lebesgue integral requirement that the derivative be bounded (Theorem 6.4.2).

Theorem 8.7.2. *If F is a differentiable function on the interval $[a, b]$, then the derivative of F, F', is H-K integrable on $[a, b]$ and*

$$\text{H-K} \int_a^x F'(t)\,dt = F(x) - F(a).$$

Proof. Let $\epsilon > 0$ be given. For each point x in $[a, b]$ we have a positive number $\delta_\epsilon(x)$ so that for $u < v$, $x - \delta_\epsilon(x) < u \leq x \leq v < x + \delta_\epsilon(x)$, and

$$\left| F(v) - F(u) - F'(x)(v - u) \right| \leq \epsilon(v - u)$$

by the Straddle Lemma (Exercise 5.9.2).

We have a gauge $\delta_\epsilon(\cdot)$. Let P be a δ_ϵ-fine partition of $[a, b]$. Then by Cousin's Lemma 8.2.1 we have

$$\{(c_1, [a, x_1]), (c_2, [x_1, x_2]), \ldots, (c_n, [x_{n-1}, b])\},$$

with

$$c_k - \delta_\epsilon(c_k) < x_{k-1} \le c_k \le x_k < c_k + \delta_\epsilon(c_k),$$

for $1 \le k \le n$. Thus

$$\left| \sum_P F'(c_k)\, \Delta x - [F(x) - F(a)] \right|$$

$$= \left| \sum_P \big(F'(c_k)(x_k - x_{k-1}) - [F(x_k) - F(x_{k-1})]\big) \right|$$

$$\le \epsilon \sum \Delta x_k = \epsilon(b - a).$$

So we have that F' is H-K integrable and H-K $\int_a^x F'(t)\, dt = F(x) - F(a)$ for x in the interval $[a, b]$. $\square$

We have shown that *every* derivative is H-K integrable, bounded or not.

8.7.3   H-K, but not Lebesgue, Integrable

We will demonstrate that an H-K function need not be Lebesgue integrable. Consider

$$F(x) = \begin{cases} x^2 \sin(\pi/x^2) & 0 < x \le 1, \\ 0 & x = 0. \end{cases}$$

Then

$$F'(x) = \begin{cases} 2x \sin(\pi/x^2) - 2\pi/x \cos(\pi/x^2) & 0 < x \le 1, \\ 0 & x = 0. \end{cases}$$

So F' is H-K integrable and H-K $\int_0^1 F'(x)\, dx = F(1) - F(0) = 0$.

However, F' is not Lebesgue integrable. Recall that a function is Lebesgue integrable iff its absolute value is Lebesgue integrable. The function $2x \sin(\pi/x^2)$ (when $0 < x \le 1$) and 0 (when $x = 0$) is continuous on $[0, 1]$.

Exercise 8.7.1. If F' was Lebesgue integrable, we could conclude that

$$\text{L} \int_0^1 \frac{2\pi}{x} \left| \cos\left(\frac{\pi}{x^2}\right) \right| dx < \infty.$$

Show that this is not the case. Hint: $x_n = \sqrt{1/(2n + 1)}$.

The H-K integral is indeed an extension of the Lebesgue integral.

Can we weaken the hypothesis that F be differentiable? The continuous function $2\sqrt{x}$ has a derivative of $1/\sqrt{x}$ for $x \neq 0$. In Example 8.3.3 the H-K integral of the function $1/\sqrt{x}$ (when $0 < x \leq 1$) and 0 (when $x = 0$) was shown to exist and to have value $2\sqrt{1}$.

Exercise 8.7.2. Show that the continuous function $F(x) = 2\sqrt{x}$, with a derivative F' defined by $1/\sqrt{x}$ for $x > 0$ and 0 when $x = 0$ satisfies $F(x) = \text{H-K} \int_0^x F'(t)\, dt$.

Maybe F does not have to be differentiable at all points. On the other hand, the Cantor function C is a continuous function, differentiable almost everywhere ($C' = 0$ almost everywhere), and (by Example 8.3.5)

$$C(1) - C(0) = 1 \neq 0 = \text{L} \int_0^1 C'\, d\mu = \text{H-K} \int_0^1 C'\, d\mu.$$

A set of measure zero, with regards to the derivative, still causes problems. It turns out we can successfully deal with a countable set of problem points. We have a Fundamental Theorem of Calculus for the H-K integral — in fact, as we shall see, we have two fundamental theorems to examine.

8.7.4   Fundamental H-K Theorem

Theorem 8.7.3 (FTC for the H-K Integral). *If F is continuous on $[a, b]$ and if F is differentiable on $[a, b]$ with at most a countable number of exceptional points, then F' is H-K integrable on $[a, b]$ and*

$$\text{H-K} \int_a^x F'(t)\, dt = F(x) - F(a) \qquad \text{for each } x \in [a, b].$$

Proof. At the exceptional points, say $a_1, a_2, \ldots, a_k, \ldots$, we define F' to be zero: $F'(a_k) \equiv 0$, with $k = 1, 2, \ldots$. Let $\epsilon > 0$ be given. We now construct a gauge on $[a, b]$.

For $x \neq a_k$, we have a positive number $\delta_\epsilon^d(x)$ so that

$$\left| F(v) - F(u) - F'(x)(v - u) \right| \leq \epsilon(v - u)$$

whenever $x - \delta_\epsilon^d(x) < u \leq x \leq v < x + \delta_\epsilon^d(x)$, by the Straddle Lemma (Exercise 5.9.2). Because F is continuous at a_k, we have a $\delta_\epsilon^c(a_k)$ so that

$$|F(x) - F(a_k)| < \frac{\epsilon}{2^{k+2}}$$

whenever $a_k - \delta_\epsilon^c(a_k) < x < a_k + \delta_\epsilon^c(a_k)$. Define the gauge $\delta_\epsilon(x)$ on $[a, b]$
by

$$\delta_\epsilon(x) = \begin{cases} \delta_\epsilon^d(x) & x \neq a_1, a_2, \ldots, \quad \text{(differentiability)} \\ \delta_\epsilon^c(a_k) & x = a_k. \quad \text{(continuity)} \end{cases}$$

This gauge will work for $[a, x]$.

By Cousin's Lemma 8.2.1 we have a δ_ϵ-fine partition of $[a, x]$:

$$\{(c_1, [a, x_1]), (c_2, [x_1, x_2]), \ldots, (c_n, [x_{n-1}, x])\}.$$

If none of the tags c_k is an exceptional point $a_1, a_2, \ldots, a_k, \ldots$, argue as
in Theorem 8.7.2. Otherwise, suppose the tag c_k is the exceptional point
a_N. If c_k tags only one subinterval, $(c_k, [x_{k-1}, x_k])$, then for $F'(c_k) = F'(a_N) = 0$ we have

$$\left| F(x_k) - F(x_{k-1}) - F'(c_k)(x_k - x_{k-1}) \right| = |F(x_k) - F(x_{k-1})|$$
$$< 2\left(\frac{\epsilon}{2^{N+2}}\right) = \frac{\epsilon}{2^{N+1}}.$$

But if c_k tags two subintervals, either $(c_k, [x_{k-2}, x_{k-1}])$ and $(c_k, [x_{k-1}, x_k])$,
or else $(c_k, [x_{k-1}, x_k])$ and $(c_k, [x_k, x_{k+1}])$, we could have

$$\left| F(x_k) - F(x_{k-1}) - F'(c_k)(x_k - x_{k-1}) \right| = |F(x_k) - F(x_{k-1})|$$

and

$$\left| F(x_{k-1}) - F(x_{k-2}) - F'(c_k)(x_{k-1} - x_{k-2}) \right| = |F(x_{k-1}) - F(x_{k-2})|$$

to estimate. In this case, we have

$$4\left(\frac{\epsilon}{2^{N+2}}\right) = \frac{\epsilon}{2^N}.$$

The worst case would be if each tag occurs as an exceptional point. But
we have only a finite number of tags, so the total contribution from such
exceptional tags must be less than $\sum \epsilon/2^N = \epsilon$.

For the tags that are not exceptional points,

$$\left| F(x_k) - F(x_{k-1}) - F'(c_k)(x_k - x_{k-1}) \right| \leq \epsilon(x_k - x_{k-1})$$

and

$$\sum \left| F(x_k) - F(x_{k-1}) - F'(c_k)(x_k - x_{k-1}) \right| \leq \epsilon(x - a) \leq \epsilon(b - a).$$

We have for any δ_ϵ-fine partition P of $[a, x]$

$$\left| F(x) - F(a) - \sum_P F'(c_k)\, \Delta x_k \right| \le \epsilon(b - a) + \epsilon.$$

Thus, F' is H-K integrable and H-K $\int_a^x F'(t)\, dt = F(x) - F(a)$. $\quad\square$

Compare Theorems 2.3.1, 3.7.1, and 6.4.2.

8.7.5   Sawtooth Functions

We offer one example of a so-called *sawtooth function*. The reader may construct other examples; these examples are informative.

We construct a function F on $[0, 1]$ as follows.

Step 1. Define F as

$$F(0) = 0, \ F\left(\frac{1}{2}\right) = 1, \ F\left(\frac{2}{3}\right) = 1 - \frac{1}{2}, \ F\left(\frac{3}{4}\right) = 1 - \frac{1}{2} + \frac{1}{3}, \ldots$$

$$F\left(\frac{n}{n+1}\right) = 1 - \frac{1}{2} + \cdots + (-1)^{n+1}\left(\frac{1}{n}\right), \ldots.$$

$$F(1) = \ln 2,$$

and let F be linear otherwise.

Step 2. F is continuous on $[0, 1]$ so $\lim_{x \to 1^-} F(x) = \ln 2$.

Step 3. F is differentiable except at $\frac{1}{2}, \frac{2}{3}, \ldots, n/(n+1), \ldots$. Thus

$$F'(x) = \begin{cases} 2 & \text{on } (0, \frac{1}{2}), \\ -3 & \text{on } (\frac{1}{2}, \frac{2}{3}), \\ 4 & \text{on } (\frac{2}{3}, \frac{3}{4}), \\ \cdots & \\ (-1)^{n+1}(n+1) & \text{on } \big((n-1)/n, n/(n+1)\big), \ldots. \end{cases}$$

Step 4. F' is H-K integrable on $[0, 1]$ and H-K $\int_0^1 F'(x)\, dx = F(1) - F(0) = \ln 2$.

Step 5. F' is not Lebesgue integrable on $[0, 1]$.

How did we reach that final step? Assume F' is Lebesgue integrable. Then $|F'|$ is Lebesgue integrable, and

$$
\begin{aligned}
\mathrm{L} \int_0^1 |F'|\, d\mu &> \mathrm{L} \int_0^{n/(n+1)} |F'|\, d\mu \\
&= 2\left(\frac{1}{2}\right) + 3\left(\frac{1}{6}\right) + 4\left(\frac{1}{12}\right) + \cdots + (n+1)\left(\frac{1}{n(n+1)}\right) \\
&= 1 + \frac{1}{2} + \frac{1}{3} + \cdots + \frac{1}{n}.
\end{aligned}
$$

We have a contradiction.

This concludes our treatment of recovering a function from its derivative using the H-K integral. How about recovering a function from its H-K integral using differentiation?

8.8   The Second Fundamental Theorem

Before proving the Second Fundamental Theorem of H-K integrals, we will discuss some properties of the function $F(x) \equiv \text{H-K} \int_a^x f(t)\, dt$, for $a \le x \le b$, where we assume that f is H-K integrable on $[a, b]$.

8.8.1   Some H-K Properties

Property 1. F is continuous on $[a, b]$.

Let $\epsilon > 0$ be given and select any point c in (a, b). We will show that F is continuous at c.

Because f is H-K integrable on $[a, b]$, we have a gauge $\delta_\epsilon(\cdot)$ so that for any δ_ϵ-fine partition of $[a, b]$,

$$
\left| \sum f(c_k)\, \Delta x - \text{H-K} \int_a^b f(t)\, dt \right| < \epsilon.
$$

Define a new gauge $\delta(\cdot)$ on $[a, b]$ by

$$
\delta(x) \equiv
\begin{cases}
\min\left\{\frac{1}{2}|x - c|, \delta_\epsilon(x)\right\} & \text{if } x \ne c, \\
\min\left\{\delta_\epsilon(c), c - a, b - c, \epsilon/(|f(c)| + 1)\right\} & \text{if } x = c.
\end{cases}
$$

For any δ-fine partition of $[a, b]$, c is a tag. For x in $(c - \delta(c), c + \delta(c))$ — that is, $|x - c| < \delta(c)$ — application of Henstock's Lemma to $(c, [x, c])$

and $(c, [c, x])$ implies

$$\left| f(c)(c - x) - \left(F(c) - F(x) \right) \right| \le \epsilon \qquad \text{and}$$
$$\left| f(c)(x - c) - \left(F(x) - F(c) \right) \right| \le \epsilon.$$

That is,

$$|F(x) - F(c)| \le \epsilon + |f(c)|\,|x - c| < \epsilon + |f(c)| \cdot \frac{\epsilon}{|f(c)| + 1} < 2\epsilon.$$

Thus, if $|x - c| < \delta(c)$, then $|F(x) - F(c)| < 2\epsilon$, so F is continous at c. The reader may consider $c = a$ and $c = b$.

Property 2. If f is nonnegative, F is nondecreasing.

For $a \le x < y \le b$, $F(y) - F(x) = \text{H-K} \int_x^y f(t)\,dt$. All the H-K sums are nonnegative.

Property 3. If f is continuous at $c \in (a, b)$, then F is differentiable at c.

Let $\epsilon > 0$ be given. Because f is continuous at c, we have a $\delta_\epsilon(c)$ so that

$$f(c) - \epsilon < f(t) < f(c) + \epsilon \qquad \text{for} \qquad |t - c| < \delta_\epsilon(c), t \in [a, b].$$

For $h > 0$,

$$h\big(f(c) - \epsilon\big) \le \text{H-K} \int_c^{c+h} f(t)\,dt \le h\big(f(c) + \epsilon\big).$$

That is,

$$-\epsilon h \le \text{H-K} \int_c^{c+h} [f(t) - f(c)]\,dt \le \epsilon h,$$

and

$$\left| \frac{F(c + h) - F(c)}{h} - f(c) \right| = \left| \text{H-K} \int_c^{c+h} \frac{[f(t) - f(c)]}{h}\,dt \right| \le \epsilon.$$

Complete the argument by considering $h < 0$, $c = a$, $c = b$, and show that

1. if $c \in [a, b)$ and $\lim_{x \to c+} f(x)$ exists, then F has a right-hand derivative at c equal to $f(c^+)$;

2. if $c \in (a, b]$ and $\lim_{x \to c^-} f(x) = f(c^-)$ exists, then F has a left-hand derivative at c equal to $f(c^-)$.

In fact, if we remove the requirement that f is continuous or has a left- or right-hand limit, it is still true that $F' = f$ almost everywhere. This is the Second Fundamental Theorem.

8.8.2   Second Fundamental Theorem for H-K

Theorem 8.8.1 (Second FTC for the H-K Integral). *Given that f is H-K integrable on the interval $[a, b]$, define a function F on $[a, b]$ by $F(x) \equiv$ H-K $\int_a^x f(t)\, dt$. Then*

1. *F is continuous on $[a, b]$,*

2. *$F' = f$ almost everywhere on $[a, b]$, and*

3. *f is Lebesgue measurable.*

Proof. We have shown that F is continuous (Section 8.8.1). To show that $F' = f$ almost everywhere on $[a, b]$, we will show that the Dini derivate $D^+ F$ equals f almost everywhere.

Let E be the set of points in $[a, b]$ for which $D^+ F$ is not f. First we will show that E has measure zero. For x in the set E,

$$\lim_{h \to 0^+} \sup \frac{F(y) - F(x)}{y - x}$$

is not f. If that upper limit is greater than $f(x)$, then for every $h > 0$ we have a number y_h^x in $[x, x + h] \cap [a, b]$, so that

$$\frac{F(y_h^x) - F(x)}{y_h^x - x} > f(x) + \eta_x,$$

with η_x a positive number. That is, $F(y_h^x) - F(x) - f(x)(y_h^x - x) > \eta_x(y_h^x - x)$.

However, if that upper limit is less than $f(x)$, then for every $h > 0$ we have a number y_h^x in $[x, x + h]$ so that

$$\frac{F(y_h^x) - F(x)}{y_h^x - x} < f(x) - \eta_x,$$

with η_x a positive number. That is, for x in the set E and every $h > 0$ we have a number y_h^x in $[x, x + h] \cap [a, b)$ so that

$$\left| F(y_h^x) - F(x) - f(x)(y_h^x - x) \right| > \eta_x(y_h^x - x).$$

For x in the set E, we have a sequence of points y_h^x in $[a, b)$ converging to x from the right so that $\left| F(y_h^x) - F(x) - f(x)(y_h^x - x) \right| > \eta_x(y_h^x - x)$, for η_x positive.

With each point x in E we associate a positive number η_x. Then

$$E = \cup \left\{ x \in E \mid \eta_x > \frac{1}{n} \right\}.$$

We will show that $E_n = \{x \in E \mid \eta_x > 1/n\}$ has measure zero for $n = 1, 2, \ldots$; thus E will have measure zero.

Fix n. Let ϵ be greater than zero. We have a gauge $\delta_\epsilon(\cdot)$ so that if P is any δ-fine partition of $[a, b]$, then

$$\left| \sum f(c)\, \Delta x_k - \text{H-K} \int_a^b f(x)\, dx \right| < \frac{\epsilon}{n}.$$

The collection of closed intervals $\{[x, y_h^x] \mid x \in E_n, x < y_h^x < x + h < x + \delta_\epsilon(x)\}$ forms a Vitali cover of E_n.

We have a finite collection $\{[x_1, y_h^{x_1}], \ldots, [x_M, y_h^{x_M}]\}$ of disjoint closed intervals, $x_1, \ldots, x_M$ points in E_n so that

$$\mu^*(E_n) - \sum_{m=1}^{M} \left(y_h^{x_m} - x_m \right) < \epsilon.$$

That is,

$$\sum_{m=1}^{M} \left(y_h^{x_m} - x_m \right) > \mu^*(E_n) - \epsilon.$$

But

$$\left[x_m, y_n^{x_m} \right] \subset \left(x_m - \delta_\epsilon(x_m), x_m + \delta_\epsilon(x_m) \right).$$

Apply Henstock's Lemma 8.5.1. We conclude

$$\frac{2\epsilon}{n} \geq \sum_{m=1}^{M} \left| f(x_m)\left(y_h^{x_m} - x_m\right) - \text{H-K} \int_{x_m}^{y_h^{x_m}} f(t)\, dt \right|$$

$$= \sum \left| f(x_m)\left(y_h^{x_m} - x_m\right) - \left[F\left(y_h^{x_m}\right) - F(x_m) \right] \right|$$

$$> \sum \eta_{x_m} \left(y_h^{x_m} - x_m \right) > \frac{1}{n}\left(\mu^*(E_n) - \epsilon \right).$$

Thus $\mu^*(E_n) < 3\epsilon$ and E_n is Lebesgue measurable since $\mu^*(E_n) = 0$.

Again, let E denote the points x in $(a, b]$ for which $F_-(x) \neq f(x)$. That is,

$$F_-(x) = \lim_{h \to 0+} \inf \left\{ \frac{F(y) - F(x)}{y - x}, x - h < y < x \right\} \neq f(x).$$

If $F_-(x) > f(x)$, then for $h > 0$ we have $y_h^x < x$ in $(a, b]$, so that

$$\frac{F\left(y_h^x\right) - F(x)}{y_h^x - x} > f(x) + \eta_x,$$

for η_x a positive number. That is,

$$F\left(y_h^x\right) - F(x) - \left(y_h^x - x\right) f(x) < \eta_x \left(y_h^x - x\right) < 0 \qquad \text{for } y_n^x - x < 0.$$

However, if $F_-(x) < f(x)$, then

$$\frac{F\left(y_h^x\right) - F(x)}{y_h^x - x} < f(x) - \eta_x.$$

That is, $F\left(y_h^x\right) - F(x) - f(x)\left(y_h^x - x\right) > -\eta_x \left(y_h^x - x\right) > 0$. In short,

$$\left| F\left(y_h^x\right) - F(x) - f(x)\left(y_h^x - x\right) \right| > -\eta_x \left(y_h^x - x\right) = \eta_x \left(x - y_h^x\right).$$

As before, $E_n = \{x \in E \mid \eta_x > 1/n\}$ and $E = \cup E_n$. Fix n. The collection of intervals $\{[y_h^x, x] \mid x \in E_n, x - \delta_\epsilon(x) < x - h < x - y_h^x < x\}$ forms a Vitali cover of E_n.

We have a finite collection $\{[y_h^{x_1}, x_1], \ldots, [y_h^{x_M}, x_M]\}$ of disjoint closed subintervals of $(a, b]$, with $\sum_{m=1}^{M} \left(x_m - y_h^{x_m}\right) > \mu^*(E_n) - \epsilon$. But

$$\left[y_h^{x_m}, x_m\right] \subset \left(x_m - \delta_\epsilon(x_m), x_m + \delta_\epsilon(x_m)\right).$$

Applying Henstock's Lemma 8.5.1, we have

$$\frac{2\epsilon}{n} \geq \sum_{m-1}^{M} \left| f(x_m)\left(x_m - y_h^{x_m}\right) - \text{H-K} \int_{y_h^{x_m}}^{x_m} f(t)\, dt \right|$$

$$= \sum_{m=1}^{M} \left| f(x_m)\left(x_m - y_h^{x_m}\right) - \left[F(x_m) - F\left(y_h^{x_m}\right)\right] \right|$$

$$> \eta_x \left(x_m - y_h^{x_m}\right) > \frac{1}{n}\left(\mu^*(E_n) - \epsilon\right).$$

Thus $\mu^*(E_n) < 3\epsilon$. By the arbitrary nature of ϵ, we have that $\mu^*(E_n) = 0$, so E_n is Lebesgue measurable and $\mu(E_n) = 0$.

Show that the other two derivates equal f almost everywhere, and consequently $F' = f$ almost everywhere.

The third conclusion — that f is Lebesgue measurable — follows, since differentiability of F almost everywhere implies that F is continuous almost everywhere and thus (Section 5.7.2) F is measurable. Define F for $x > b$ by $F(x) = F(b)$. Then $f_n(x) = n\,[F(x + (1/n)) - F(x)]$ is a measurable function. Thus, $\lim f_n(x) = F'(x)$ almost everywhere on $[a, b]$ is a measurable function (Theorem 5.7.1). Since $F' = f$ almost everywhere, f is measurable (Section 5.7.2). $\quad\square$

Compare Theorems 2.4.1, 3.7.2, and 6.4.1. This concludes our discussion of the H-K Integral.

8.9 Summary

Two Fundamental Theorems of Calculus for the Henstock–Kurzweil Integral

If F is continuous on $[a, b]$ and F is differentiable with at most a countable number of exceptions on $[a, b]$, then

1. F' is Henstock–Kurzweil integrable on $[a, b]$ and

2. H-K $\int_a^x F'(t)\,dt = F(x) - F(a)$, $a \le x \le b$.

If f is Henstock–Kurzweil integrable on $[a, b]$ and $F(x) = $ H-K $\int_a^x f(t)\,dt$, then

1. F is continuous on $[a, b]$,

2. $F' = f$ almost everywhere on $[a, b]$, and

3. f is measurable.

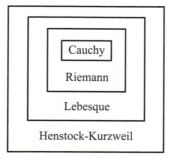

Figure 2. C $\subset$ R $\subset$ L $\subset$ H-K integrable functions.

8.10   References

1. Bartle, Robert. *A Modern Theory of Integration.* Graduate Studies in Mathematics, Vol. 32. Providence, R.I.: American Mathematical Society, 2001.

2. DePree, John, and Charles Schwartz. *Introduction to Analysis.* New York: Wiley, 1988.

3. Gordon, Russell. *The Integrals of Lebesgue, Denjoy, Perron, and Henstock.* Graduate Studies in Mathematics, Vol. 4. Providence, R.I.: American Mathematical Society, 1994.

4. McLeod, Robert. *The Generalized Riemann Integral.* Carus Monographs, No. 20. Washington: Mathematical Association of America, 1980.

5. Yee, Lee, and Rudolf Vyborny. *The Integral: An Easy Approach, after Kurzweil and Henstock.* Cambridge University Press, 2000.

CHAPTER **9**

The Wiener Integral

What science can there be more noble, more excellent, more useful for men, more admirably high and demonstrative than this of mathematics?
— Benjamin Franklin

In the preceding chapters, the integrals under discussion were defined on sets of real numbers. So the domains of integration have consisted of real numbers. In contrast, the Wiener integral has as its domain of integration *the space of continuous functions on the interval* [0, 1] *that begin at the origin.* A continuous function now plays the role of a real number.

With the Wiener integral *path* replaces *point*: these are integrals over sets of continuous functions, integrals over "paths." Hence the terminology *path integral*. The approach, as in the development of the Lebesgue integral, is threefold:

1. We begin by defining a measure on special subsets of our space of continuous functions.

2. We extend this measure to an appropriate sigma algebra.

3. With a measure in place, we develop an integration process leading to the Wiener integral.

9.1 Brownian Motion

The story begins in Scotland in the 1820s with a botanist named Robert Brown (discoverer of nuclei of plant cells) who was studying the erratic motion of organic and inorganic particles (pollen and ground silica, respectively) suspended in liquid. Neighboring particles experienced unrelated motion, movement was equally likely in any direction, and past motion had no bearing on future motion.

Despite numerous experiments that varied heat and viscosity, a satisfactory explanation was not forthcoming. Decades later a theoretical explanation was put forward by Albert Einstein (1905) and Marian Von Smoluchowski (1906); the eventual experimental verification by Jean Perrin (1909) resulted in his receiving the 1926 Nobel prize in physics.

Consider a collection of particles with initial displacement zero. Einstein argued that the number of particles per unit volume around position x at time t (in other words, the particle density at x at time t) is given by $(2\pi t)^{-1/2}e^{-x^2/2t}$, which is a solution of the diffusion equation. Moreover, the mean square displacement from the beginning position is proportional to the elapsed time t.

Consider a single particle, replacing *density* by *probability*. Starting from position $x = 0$ at time $t = 0$, the probability that the particle will be found between a and b at time t is written $P\big(a \leq x(t) < b \mid x(0) = 0\big)$. We calculate this probability by

$$P\big(a \leq x(t) < b \mid x(0) = 0\big) = (2\pi t)^{-1/2} \int_a^b e^{-x^2/2t}\, dx. \qquad (1)$$

The mean square displacement — the *expected value* of x^2 — may be calculated by

$$E(x^2) = (2\pi t)^{-1/2} \int_{-\infty}^{\infty} x^2 e^{-x^2/2t}\, dx = t,$$

and the particle density, $(2\pi t)^{-1/2}e^{-x^2/2t}$, is a solution of the diffusion equation

$$\frac{\partial u}{\partial t} = \frac{1}{2}\frac{\partial^2 u}{\partial x^2}.$$

Thinking of a particle starting at the origin at time zero, we are assessing its chances, its probability, of passing through the *window* $[a, b)$ at time t.

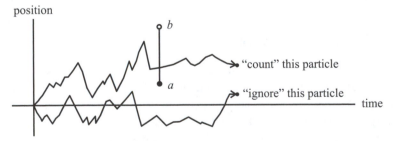

Figure 1. Particle probability

Exercise 9.1.1.

a. Using equation 1, calculate numerically the probability that a particle at time $t = \frac{1}{2}$ passes through the windows $[-1, 1)$, $[-2, 2)$, $[-3, 3)$, assuming $x(0) = 0$. Hint: As the window increases in size, we should count more particles, and for the largest window $[-\infty, \infty)$, we should count all particles.

b. Show $P\left(-\infty \leq x(t) < \infty \mid x(0) = 0\right) = (2\pi t)^{-1/2} \int_{-\infty}^{\infty} e^{-x^2/2t} \, dx$ $= 1$, for $t > 0$.

We would expect the particle to wander up or down (so to speak) with equal probability. That is, the expected space coordinate x at time t, for a particle beginning at the origin, should be zero. Readers with some familiarity with probability will understand that the expected value of x, $E(x)$, should be zero.

Exercise 9.1.2.

a. Show $E(x) = (2\pi t)^{-1/2} \int_{-\infty}^{\infty} x e^{-x^2/2t} \, dx = 0$. Hint: Symmetry or integration by parts.

b. Show $E(x) = \left(2\pi(t - t_1)\right)^{-1/2} \int_{-\infty}^{\infty} x e^{-(x-\xi_1)^2/2(t-t_1)} \, dx = \xi_1$, for $t > t_1$, assuming $x(t_1) = \xi_1$. Hint: If the particle is at location ξ_1 at t_1, its expected location $t - t_1$ seconds later is ξ_1. There is nothing special about starting at $x = 0$ at $t = 0$.

Here is another thought about the previous exercise. If we started with a collection of particles at the origin at time zero, we would expect the particles to spread out — to disperse away from zero — although most should remain close to zero, the expected value. The reader may recall that the variance is a measure of the dispersion away from the expected value; in this setting, it's away from zero. Formally, the variance of position x, written as $E\left((x - E(x))^2\right)$, equals $E(x^2)$ since $E(x) = 0$.

Exercise 9.1.3.

a. Show $E(x^2) = (2\pi t)^{-1/2} \int_{-\infty}^{\infty} x^2 e^{-x^2/2t} \, dx = t$.

b. Assuming $x(t_1) = \xi_1$, show $E(x^2) = t - t_1$. Hint: The expected value of the square of the displacement, the so-called *mean square displacement*, is proportional to the elapsed time.

Exercise 9.1.4. Calculate numerically

$$P\left(-\frac{1}{2} \le x\left(\frac{1}{4}\right) < \frac{1}{2} \mid x(0) = 0\right),$$

$$P\left(-\frac{1}{\sqrt{2}} \le x\left(\frac{1}{2}\right) < \frac{1}{\sqrt{2}} \mid x(0) = 0\right),$$

$$P\left(-\frac{\sqrt{3}}{2} \le x\left(\frac{3}{4}\right) < \frac{\sqrt{3}}{2} \mid x(0) = 0\right), \text{ and}$$

$$P\left(-1 \le x(1) < 1 \mid x(0) = 0\right).$$

Exercise 9.1.5. Show that the function $(2\pi t)^{-1/2}e^{-x^2/2t}$ is a solution of the diffusion equation $u_t = \frac{1}{2}u_{xx}$.

Yes, the distribution of the particles' position, the distribution of x at time t, starting from the origin $(x(0) = 0)$, is a normal distribution with expected value 0 and variance t.

9.1.1 Wiener's Explanation

In the 1920s Norbert Wiener developed a mathematical explanation of Brownian motion, a staggering achievement given the complexity of the physical phenomenon. Wiener wrote: "In the Brownian movement... it is the displacement of a particle over one interval that is independent of the displacement of the particle over another interval" (*Collected Works*, 1976, p. 459).

That is, the quantities (random variables)

$$x(t_1) - x(0), x(t_2) - x(t_1), \ldots, x(t_n) - x(t_{n-1})$$

vary independently, $0 < t_1 < t_2 < \cdots < t_n \le 1$, and are normally distributed with mean zero and variance $t_k - t_{k-1}$, with $k = 1, 2, \ldots, n$, respectively.

We have this mathematical model of Brownian motion:

1. $x(0) = 0$ (all particles begin at the origin).

2. $x(\cdot)$ is continuous for $0 \le t \le 1$ (erratic, but continuous paths).

3. The random variable $x(t) - x(s)$, which is the change in position over the time interval $t - s$, has a normal distribution with mean zero and variance $t - s$, for $0 < s < t \le 1$. We write

$$P\left(a \le x(t) - x(s) < b\right) = \left(2\pi(t - s)\right)^{-1/2} \int_a^b e^{-\xi^2/2(t-s)} d\xi.$$

4. $x(t_1) - x(0), \ldots, x(t_n) - x(t_{n-1})$ are independent random variables for $0 < t_1 < \cdots < t_n \leq 1$.

From these intuitive concepts we now develop Wiener measure. Note that in what follows

$$K(x,t) = (2\pi t)^{-1/2} \exp\left(-\frac{x^2}{2t}\right), \qquad \text{for } 0 < t \leq 1.$$

9.2 Construction of the Wiener Measure

Our construction of Lebesgue measure on the space of real numbers R^1 began by assigning a measure (length) to intervals of real numbers: $\ell\big([a,b)\big) \equiv b - a$. For Wiener measure, our space will be the continuous functions on $[0,1]$ beginning at the origin, C_0, with the *sup norm*:

$$\|x\| = \max_{0 \leq t \leq 1} |x(t)|.$$

The norm defines a topology, sometimes called the metric topology on C_0. A set $\{x \in C_0 | \|x - x_0\| < r\}$ is called an open ball with center x_0 and radius r. A subset E of C_0 is called an open set if for each x in E there exists $r > 0$ so the $\{y | \|y - x\| < r\} \subset E$. In Section 9.3 we develop these ideas.

When we assign a measure to the "quasi-interval" of continuous functions, the set

$$\{x(\cdot) \in C_0 \mid a_1 \leq x(t_1) < b_1, \, 0 < t_1 \leq 1\},$$

what do we have? It is the probability of a Brownian particle starting at the origin and passing through the window $[a_1, b_1)$ at time t_1, where the position at time t_1, $x(t_1)$, satisfies $a_1 \leq x(t_1) < b_1$. Figure 2 illustrates this probability. We write

$$w_{t_1}\big(\{x(\cdot) \in C_0 \mid a_1 \leq x(t_1) < b_1, \, 0 < t_1 \leq 1\}\big)$$

$$\equiv \mathrm{L} \int_{a_1}^{b_1} (2\pi t_1)^{-1/2} \exp\left(-\frac{\xi_1^2}{2t_1}\right) d\xi_1.$$

9.2.1 Borel Cylinders with One Restriction

More generally, we may replace the interval $[a_1, b_1)$ with any Borel set B^1 in R^1, $d\xi_1$ denoting Lebesgue measure in R^1; see Figure 3. We write

$$\mathcal{B}_{t_1} = \big\{x(\cdot) \in C_0 \mid x(t_1) \in B^1, \, B^1 \text{ a Borel set in } \mathrm{R}^1, 0 < t_1 \leq 1\big\}$$

position

Figure 2. The window $[a_1, b_1)$

and

$$w_{t_1}(\mathcal{B}_{t_1}) = L \int_{B^1} d\xi_1 K(\xi_1, t_1).$$

The set of continuous functions $\mathcal{B}_{t_1}$ is called a *Borel cylinder with one restriction*.

Exercise 9.2.1. Fix $0 < t_1 \leq 1$. The collection of all Borel sets in R^1 is a sigma algebra of subsets of R^1, $\mathcal{B}^1$. To each Borel set B^1 in R^1, we correspond the Borel cylinder in C_0,

$$\mathcal{B}_{t_1} = \left\{ x(\cdot) \in C_0 \mid x(t_1) \in B^1 \right\}.$$

Show that this collection of Borel cylinders is a sigma algebra of subsets of C_0, $\mathcal{B}_{t_1}^1$, with $w_{t_1}(\mathcal{B}_{t_1}) = L \int_{B^1} d\xi_1 K(\xi_1, t_1)$ as the measure; $d\xi_1$ is Lebesgue measure.

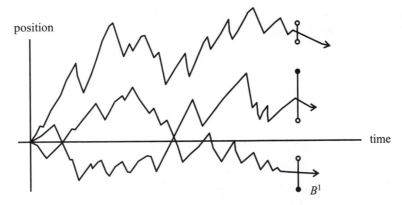

Figure 3. Borel cylinder, one restriction

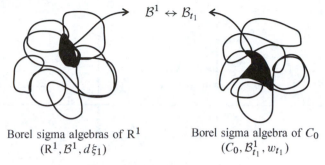

Borel sigma algebras of R^1 Borel sigma algebra of C_0
$(R^1, \mathcal{B}^1, d\xi_1)$ $(C_0, \mathcal{B}^1_{t_1}, w_{t_1})$

Figure 4. Borel sigma algebras

9.2.2 Borel Cylinders with Two Restrictions

Now suppose $0 < t_1 < t_2 \le 1$. Consider the quasi-interval of continuous functions

$$\{x(\cdot) \in C_0 \mid a_1 \le x(t_1) < b_1,\ a_2 \le x(t_2) < b_2\}.$$

(Think of Brownian particles beginning at the origin and passing through the windows $[a_i, b_i)$ at times t_i.) This situation is depicted in Figure 5.

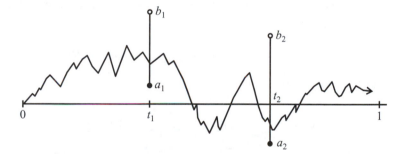

Figure 5. Borel cylinder, two restrictions

Assign a w-measure of

$$L \int_{a_2}^{b_2} d\xi_2 \int_{a_1}^{b_1} d\xi_1 K(\xi_1, t_1) K(\xi_2 - \xi_1, t_2 - t_1).$$

Why this measure? It is determined as follows.

Given the probability of a Brownian particle starting at the origin at time zero and passing through the window $[a_1, b_1)$ at time t_1 and given, furthermore, that it has passed through this window at time t_1, let us calculate its probability of passing through the window $[a_2, b_2)$ at time $t_2 > t_1$.

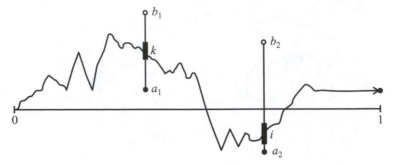

Figure 6. The kith subrectangle

First, we partition $[a_1, b_1) \times [a_2, b_2)$ into $(2^n)^2$ subrectangles. We write

$$\frac{b_1 - a_1}{2^n} \cdot \frac{b_2 - a_2}{2^n} = \Delta_1 \cdot \Delta_2.$$

Now, what is an appropriate contribution for the kith subrectangle? See Figure 6.

The probability of the particle passing through the kth segment at time t_1 is approximately $K(a_1 + k\Delta_1, t_1)\Delta_1$. Given that the particle is at the kth segment at time t_1, the probability that it is at the ith segment at time t_2 is approximately

$$K(a_1 + k\Delta_1, t_1)K(a_2 + i\Delta_2 - a_1 - k\Delta_1, t_2 - t_1)\Delta_2.$$

Thus the total contribution of that kth segment would be

$$K(a_1 + k\Delta_1, t_1)\Delta_1 \sum_{i=0}^{2^n-1} K(a_2 + i\Delta_2 - a_1 - k\Delta_1, t_2 - t_1)\Delta_2.$$

Summing over k, we have

$$\sum_{k=0}^{2^n-1}\sum_{i=0}^{2^n-1} K(a_1 + k\Delta_1, t_1)\Delta_1 K(a_2 + i\Delta_2 - a_1 - k\Delta_1, t_2 - t_1)\Delta_2$$

$$\approx \mathrm{L} \int_{a_2}^{b_2} d\xi_2 \int_{a_1}^{b_1} d\xi_1 K(\xi_1, t_1)K(\xi_2 - \xi_1, t_2 - t_1).$$

In Figure 7 the contribution is $\hat{\xi}_1 = a_1 + k\Delta_1$, $\hat{\xi}_2 = a_2 + i\Delta_2$:

$$K\left(\hat{\xi}_1, t_1\right) K\left(\hat{\xi}_2 - \hat{\xi}_1, t_2 - t_1\right) \Delta\hat{\xi}_1 \Delta\hat{\xi}_2.$$

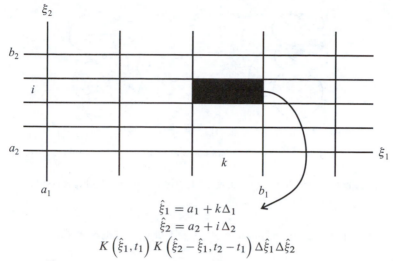

$$\hat{\xi}_1 = a_1 + k\Delta_1$$
$$\hat{\xi}_2 = a_2 + i\Delta_2$$
$$K\left(\hat{\xi}_1, t_1\right) K\left(\hat{\xi}_2 - \hat{\xi}_1, t_2 - t_1\right) \Delta\hat{\xi}_1 \Delta\hat{\xi}_2$$

Figure 7. Shaded area is the contribution of the kith segment

Next, as before, we replace $[a_1, b_1) \times [a_2, b_2)$ with an arbitrary Borel set B^2 in R^2. That is, for the Borel cylinder

$$\mathcal{B}_{t_1 t_2} = \left\{x(\cdot) \in C_0 \mid \left(x(t_1), x(t_2)\right) \in B^2\right\}$$

we have the measure

$$w_{t_1 t_2}(\mathcal{B}_{t_1 t_2}) = \mathrm{L}\int\int_{B^2} K(\xi_1, t_1)K(\xi_2 - \xi_1, t_2 - t_1)d(\xi_1, \xi_2),$$

where $d(\xi_1, \xi_2)$ denotes Lebesgue measure in R^2. The set of continuous functions $\mathcal{B}_{t_1 t_2}$ is a *Borel cylinder with two restrictions*.

Example 9.2.1. Fix $0 < t_1 < t_2 \le 1$. The collection of all Borel cylinders in C_0 whose base is a Borel set in R^2 is a sigma algebra of subsets of C_0.

The procedure is clear. Fix $0 < t_1 < t_2 < \cdots < t_n \le 1$. To a Borel set B^n in R^n, correspond the Borel cylinder $\mathcal{B}_{t_1 t_2 \ldots t_n}$ in C_0. We have

$$\left\{x(\cdot) \in C_0 \mid \left(x(t_1), \ldots, x(t_n)\right) \in B^n\right\},$$

and we assign a measure,

$$w_{t_1 t_2 \ldots t_n}(\mathcal{B}_{t_1 t_2 \ldots t_n}) = \mathrm{L}\int\cdots\int_{B^n} K(\xi_1, t_1)K(\xi_2 - \xi_1, t_2 - t_1)$$
$$\cdots K(\xi_n - \xi_{n-1}, t_n - t_{n-1})d(\xi_1, \xi_2, \ldots, \xi_n).$$

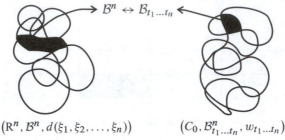

$$\mathcal{B}^n \leftrightarrow \mathcal{B}_{t_1 \ldots t_n}$$

$$\left(\mathrm{R}^n, \mathcal{B}^n, d(\xi_1, \xi_2, \ldots, \xi_n)\right) \qquad \left(C_0, \mathcal{B}^n_{t_1 \ldots t_n}, w_{t_1 \ldots t_n}\right)$$

Figure 8. Borel cylinders and sigma algebras of subsets

This collection of Borel cylinders in C_0 is a sigma algebra of subsets of C_0, $\mathcal{B}^n_{t_1 t_2 \ldots t_n}$; see Figure 8.

Let's calculate the w-measure of some Borel cylinders in C_0.

Exercise 9.2.2.

a. Suppose

$$\mathcal{B}_{t_1} = \{x(\cdot) \in C_0 \mid x(t_1) \in \mathrm{R}^1, \ 0 < t_1 \le 1\} \qquad \text{and}$$
$$\mathcal{B}_{t_1 t_2} = \{x(\cdot) \in C_0 \mid (x(t_1), x(t_2)) \in \mathrm{R}^2, \ 0 < t_1 < t_2 \le 1\}.$$

Since $C_0 = \mathcal{B}_{t_1} = \mathcal{B}_{t_1 t_2}$, we want $w_{t_1}(\mathcal{B}_{t_1}) = w_{t_1 t_2}(\mathcal{B}_{t_1 t_2}) = 1$. Show that this is in fact the case.

b. Show that for $0 < t_1 < t_2 < t_3 \le 1$,

$$w_{t_1 t_2 t_3}\left(\ \{x(\cdot) \in C_0 \mid (x(t_1), x(t_2), x(t_3)) \in [a_1, b_1) \times \mathrm{R}^1 \times [a_3, b_3)\}\ \right)$$
$$= w_{t_1 t_3}\left(\ \{x(\cdot) \in C_0 \mid (x(t_1), x(t_3)) \in [a_1, b_1) \times [a_3, b_3)\}\ \right).$$

Hint: Fubini's Theorem.

c. Show that for $0 < t_1 < t_2 < t_3 \le 1$,

$$\mathrm{L}\int_{\mathrm{R}^1} K(\xi_2 - \xi_1, t_2 - t_1) \cdot K(\xi_3 - \xi_2, t_3 - t_2) d\xi_2 = K(\xi_3 - \xi_1, t_3 - t_1)$$

(Chapman–Kolmogorov Equation).

9.2.3 The Sigma Algebra $\mathcal{B}^{[0,1]}$

What are we to do with all these collections of Borel cylinders and sigma algebras? Suppose we form the collection of all Borel cylinders with one restriction, $\mathcal{B}_{t_1}$, as t_1 varies through $(0, 1]$, a collection of sigma algebras of subsets of C_0.

Now form the collection of all Borel cylinders with two restrictions, t_1, t_2, where $0 < t_1 \neq t_2 \leq 1$. This gives us a collection of sigma algebras that contains the sigma algebras of one restriction.

Continuing in this way, form the collection of all such sigma algebras with n restrictions, $t_1, t_2, \ldots, t_n$, and so forth.

When we gather together all these collections, we have a large collection of finitely generated sigma algebras of subsets of C_0 and the associated probability measures. Let $\mathcal{B}^{[0,1]}$ denote the smallest sigma algebra generated by this collection of finitely generated sigma algebras. This large collection of sigma algebras has now been replaced by one sigma algebra, $\mathcal{B}^{[0,1]}$.

Can we similarly replace the large collection of associated probability measures with one probability measure on $\mathcal{B}^{[0,1]}$? That is what Wiener accomplished — a measure on $C_0[0, 1]$, a measure on an infinite-dimensional space.

9.3   Wiener's Theorem

Wiener takes us from the space of continuous functions to a probability space.

Theorem 9.3.1 (Wiener, 1922). *Let C_0 denote the space of continuous functions $x(\cdot)$ on $[0, 1]$ with $x(0) = 0$ and $\|x\| = \max_{0 \leq t \leq 1} |x(t)|$. For $0 < t_1 < t_2 < \cdots < t_n \leq 1$ and B^n a Borel set in $\mathbf{R}^n$, form the Borel cylinder*

$$\mathcal{B}_{t_1 t_2 \cdots t_n} = \left\{ x(\cdot) \in C_0 \mid \left(x(t_1), \ldots, x(t_n) \right) \in B^n \right\}$$

and its associated probability measure $w_{t_1 t_2 \ldots t_n}$:

$$w_{t_1 t_2 \ldots t_n} \left(\mathcal{B}_{t_1 t_2 \ldots t_n} \right) = \mathrm{L} \int \cdots \int_{B^n} K(\xi_1, t_1)$$
$$\cdots K(\xi_n - \xi_{n-1}, t_n - t_{n-1}) d(\xi_1, \xi_2, \ldots, \xi_n).$$

These probability measures may be extended uniquely to a probability measure, the Wiener measure μ_w, on the sigma algebra generated by the collection of all finitely restricted sigma algebras of subsets of C_0, $\mathcal{B}^{[0,1]}$.

A proof may be found in Yeh (1973).

What kinds of sets are in $\mathcal{B}^{[0,1]}$? Recall that with Lebesgue measure μ on R, Lebesgue measurable sets E may be covered "tightly" by open sets

G, where $\mu(G - E) < \epsilon$. They may also be approximated tightly from the inside by closed sets F, where $\mu(E - F) < \epsilon$.

In fact, we have Borel sets so that

$$\cup F_k \subset E \subset \cap G_k \qquad \text{and} \qquad \mu(\cup F_k) = \mu(E) = \mu(\cap G_k),$$

with F_k closed and G_k open. Borel sets and Lebesgue measurable sets are closely related.

Do we have similar results for Wiener measure? We will show that $\mathcal{B}^{[0,1]} = \mathcal{B}(C_0)$, where $\mathcal{B}(C_0)$ denotes the sigma algebra of Borel sets in the topological space C_0 with metric the sup norm.

Example 9.3.1. Note that

$$C_0 = \{x(\cdot) \in C_0 \mid x(t_1) \in \mathrm{R}^1, \ 0 < t_1 \leq 1\} \qquad \text{and}$$
$$\phi = \{x(\cdot) \in C_0 \mid 0 \leq x(t_1) < 0\}$$

are members of $\mathcal{B}^{[0,1]}$. Consider the functions of C_0 satisfying $|x(t)| \leq \beta$, for $0 \leq t \leq 1$ and $\beta > 0$: $\{x(\cdot) \in C_0 \mid -\beta \leq x(t) \leq \beta, \ 0 \leq t \leq 1\}$. We will show that this subset of C_0 is actually a member of $\mathcal{B}^{0,1]}$, even though it has an uncountable number of t restrictions.

This appears to be a Borel cylinder — $(-\beta, \beta)$ is a Borel set in R^1 — but we have an uncountable number of t restrictions. Select a countable dense subset $\{t_1, t_2, \ldots\}$ of $[0, 1]$, and define a sequence $\{S_n\}$ of Borel cylinders of C_0 as follows:

$$S_1 = \{x(\cdot) \in C_0 \mid -\beta \leq x(t_1) \leq \beta\},$$
$$S_2 = \{x(\cdot) \in C_0 \mid -\beta \leq x(t_1) \leq \beta, \ -\beta \leq x(t_2) \leq \beta\},$$
$$\vdots$$
$$S_n = \{x(\cdot) \in C_0 \mid -\beta \leq x(t_k) \leq \beta, \ k = 1, 2, \ldots, n\},$$
$$\vdots$$

Clearly S_n is a Borel cylinder in C_0, a member of $\mathcal{B}^{[0,1]}$, and $S_1 \supset S_2 \supset \cdots \supset S_n \supset \ldots$.

We claim that

$$S_n = \bigcap_{m=1}^{\infty} \left\{ x(\cdot) \in C_0 \mid -\beta - \frac{1}{m} < x(t_k) \leq \beta, \ k = 1, 2, \ldots, n \right\}$$

$$= \bigcap_{m=1}^{\infty} \left\{ x(\cdot) \in C_0 \mid -\beta \leq x(t_k) < \beta + \frac{1}{m}, \ k = 1, 2, \ldots, n \right\}.$$

If $\hat{x} \in S_n$, then $\hat{x}$ is continuous and $-\beta \leq \hat{x}(t_k) \leq \beta$, where $k = 1, 2, \ldots, n$. Thus we have $-\beta - 1/m < -\beta \leq \hat{x}(t_k) \leq \beta$ for all m and $k = 1, 2, \ldots, n$. That is,

$$\hat{x} \in \bigcap_{m=1}^{\infty} \left\{ x(\cdot) \in C_0 \mid -\beta - \frac{1}{m} < x(t_k) \leq \beta, \; ; k = 1, 2, \ldots, n \right\}.$$

Assume

$$\hat{x} \in \bigcap_{m=1}^{\infty} \left\{ x(\cdot) \in C_0 \mid -\beta - \frac{1}{m} < x(t_k) \leq \beta, \; ; k = 1, 2, \ldots, n \right\}.$$

Then $\hat{x}$ belongs to C_0, and $-\beta - 1/m < \hat{x}(t_k) \leq \beta$, for $k = 1, 2, \ldots, n$. We have that $\hat{x}$ belongs to C_0, and $-\beta \leq \hat{x}(t_k) \leq \beta$, for $k = 1, 2, \ldots, n$. That is, $\hat{x}$ is a member of S_n.

As a countable intersection of members of $\mathcal{B}^{[0,1]}$, S_n is a member of $\mathcal{B}^{[0,1]}$. We claim that

$$\{ x(\cdot) \in C_0 \mid -\beta \leq x(t) \leq \beta, \, 0 \leq t \leq 1 \} = \bigcap_{n=1}^{\infty} S_n.$$

Certainly this set is a subset of S_n, where $n = 1, 2, \ldots$. So it is a subset of $\cap_{n=1}^{\infty} S_n$.

Assume $\hat{x} \in \cap_{n=1}^{\infty} S_n$. Then $\hat{x}$ is continuous on $[0, 1]$, and $-\beta \leq \hat{x}(t_i) \leq \beta$, for $i = 1, 2, \ldots$ and $-\beta \leq \hat{x}(\cdot) \leq \beta$, on a dense subset of $[0, 1]$.

On the other hand, if $\hat{x} \notin \{ x(\cdot) \in C_0 \mid -\beta \leq x(t) \leq \beta, \, 0 \leq t \leq 1 \}$, where are we? Because $\hat{x}$ is continuous, we have a point t_0 in $(0, 1]$ with $\hat{x}(t_0) > \beta + \alpha$ or $\hat{x}(t_0) < -\beta - \alpha$ for some $\alpha > 0$.

Furthermore, we have a sequence $\{t_{n_k}\}$ from our dense subset of $[0, 1]$, with $t_{n_k} \to t_0$. By continuity of $\hat{x}$, we have $\hat{x}(t_{n_k}) \to \hat{x}(t_0)$. That is, $\left| \hat{x}(t_{n_k}) \right| > \beta$.

For $M > n_k$, $\hat{x} \notin S_M$. Consequently $\hat{x} \notin \cap_{n=1}^{\infty} S_n$. Thus

$$\hat{x} \in \{ x(\cdot) \in C_0 \mid -\beta \leq x(t) \leq \beta, \, 0 \leq t \leq 1 \}.$$

We conclude that $\{ x(\cdot) \in C_0 \mid -\beta \leq x(t) \leq \beta, \, 0 \leq t \leq 1 \}$ belongs to $\mathcal{B}^{[0,1]}$. The so-called closed ball center $x \equiv 0$ and radius β. We have sets with an uncountable number of restrictions, in the finitely generated sigma algebra $\mathcal{B}^{[0,1]}$.

Example 9.3.2. We claim that the set $\{x(\cdot) \in C_0 \mid |x| < \beta\}$ belongs to $\mathcal{B}^{[0,1]}$. If we can show that

$$\{x(\cdot) \in C_0 \mid |x| < \beta\} = \bigcup_{\substack{n \\ 1/n < \beta}} \left\{x(\cdot) \in C_0 \mid |x| \leq \beta - \frac{1}{n}\right\},$$

then the argument would be complete, since

$$\{x(\cdot) \in C_0 \mid |x| \leq \beta - 1/n, \ 0 \leq t \leq 1\}$$

belongs to $\mathcal{B}^{[0,1]}$, by the previous example.

Suppose $\hat{x} \in \{x(\cdot) \in C_0 \mid |x| < \beta\}$. Then $|\hat{x}| < \beta$ for all t in $[0, 1]$. Since $\hat{x}$ is continous and it assumes maximum and minimum values on $[0, 1]$, we have $-\beta < \hat{x}(\hat{t}_m) \leq \hat{x}(t) \leq \hat{x}(\hat{t}_M) < \beta$. That is, we have a natural number M so that $\beta > 1/M$ and

$$-\beta + \frac{1}{M} \leq \hat{x}(t) \leq \beta - \frac{1}{M}, \qquad \text{for } 0 \leq t \leq 1.$$

Put another way,

$$\hat{x} \in \left\{x(\cdot) \in C_0 \mid |x| \leq \beta - \frac{1}{M}\right\}.$$

Clearly,

$$\left\{x(\cdot) \in C_0 \mid |x| \leq \beta - \frac{1}{n}\right\} \subseteq \{x(\cdot) \in C_0 \mid |x| \leq \beta\}.$$

Exercise 9.3.1. Show the set $\{x(\cdot) \in C_0 \mid |x(t) - x_0(t)| < \epsilon, \ 0 \leq t \leq 1\}$ belongs to $\mathcal{B}^{[0,1]}$, for x_0 any member of C_0. Hint:

$$\{x(\cdot) \in C_0 \mid |x(t) - x_0(t)| < \epsilon, \ 0 \leq t \leq 1\}$$

$$= \bigcup_n \left\{x(\cdot) \in C_0 \mid |x(t) - x_0(t)| \leq \epsilon - \frac{1}{n}\right\}$$

$$= \bigcup_n \bigcap_k \left\{x(\cdot) \in C_0 \mid |x(t_k) - x_0(t_k)| \leq \epsilon - \frac{1}{n}\right\}.$$

The reader may conclude that

$$\{x(\cdot) \in C_0 \mid \|x\| \leq \beta\}, \ \{x(\cdot) \in C_0 \mid \|x\| < \beta\},$$

$$\text{and} \quad \{x(\cdot) \in C_0 \mid \|x - x_0\| < \epsilon\}$$

are members of $\mathcal{B}^{[0,1]}$. (Recall $\|x\| = \max_{0 \leq t \leq 1} |x(t)|$.)

The set $\{x(\cdot) \in C_0 \mid \|x - x_0\| < \epsilon\}$ is called the open ball with center x_0 of radius ϵ.

9.3.1 Open Sets and Open Balls

Every open set in a separable metric space is a countable union of open balls. We have shown the open ball in C_0 with center x_0 and radius ϵ, $\{x(\cdot) \in C_0 \mid \|x - x_0\| < \epsilon\}$, is a member of $\mathcal{B}^{[0,1]}$. If we can show that C_0 with the sup metric is a separable metric space, we may conclude that every open set in C_0 is a member of $\mathcal{B}^{[0,1]}$. That is, we may conclude that $\mathcal{B}(C_0) \subset \mathcal{B}^{[0,1]}$.

Example 9.3.3. Form the collection of all polygonal functions on $[0, 1]$. These are linear on $[(k-1)/n, k/n]$, for $k = 1, 2, \ldots, n$, vanish at $t = 0$, and assume rational values at k/n. We will show that this countable collection of "points" is a dense subset of C_0.

Pick an arbitrary open ball in C_0, $\{x(\cdot) \in C \mid \|x - x_0\| < \epsilon\}$. Now construct a polygonal function $\hat{x}$ belonging to this ball as follows. Let $\epsilon > 0$ be given. Because x_0 is uniformly continuous on $[0, 1]$, there exists $\delta > 0$ so that $|x_0(t) - x(s)| < \epsilon/5$ whenever $|t - s| < \delta$ for all t, x in $[0, 1]$. Choose n so that $1/n < \delta$.

We have $0 < 1/n < 2/n < \cdots < n/n = 1$. So $|x_0(u) - x_0(v)| < \epsilon/5$ whenever we have $u, v \in [(k-1)/n, k/n]$, for $k = 1, 2, \ldots, n$.

In each subinterval $((k-1)/n, k/n]$, pick a rational number r_k and define the polygonal function $\hat{x}$ so that $\hat{x}(k/n) = x_0(r_k)$, with $\hat{x}(0) = 0$.

For $t \in [(k-1)/n, k/n]$, we calculate

$$
\begin{aligned}
|\hat{x}(t) - x_0(t)| &= \left| \hat{x}(t) - \hat{x}\left(\frac{k-1}{n}\right) \right| + \left| \hat{x}\left(\frac{k-1}{n}\right) - x_0\left(\frac{k-1}{n}\right) \right| \\
&\quad + \left| x_0\left(\frac{k-1}{n}\right) - x_0(t) \right| \\
&= |\hat{x}(t) - x_0(r_{k-1})| + \left| x_0(r_{k-1}) - x_0\left(\frac{k-1}{n}\right) \right| \\
&\quad + \left| x_0\left(\frac{k-1}{n}\right) - x_0(t) \right| \\
&\leq |\hat{x}(t) - x_0(r_{k-1})| + \frac{2\epsilon}{5} \\
&\leq \left| \hat{x}(t) - x_0\left(\frac{k}{n}\right) \right| + \left| x_0\left(\frac{k}{n}\right) - x_0\left(\frac{k-1}{n}\right) \right| \\
&\quad + \left| x_0\left(\frac{k-1}{n}\right) - x_0(r_{k-1}) \right| + \frac{2\epsilon}{5} < \epsilon.
\end{aligned}
$$

Thus $\|\hat{x} - x_0\| < \epsilon$.

We have shown that every open ball contains a "polygonal function." Suppose G is an open set in C_0 and x_0 belongs to G. We have $\epsilon > 0$ so that

$$\{x(\cdot) \in C_0 \mid \|x - x_0\| < \epsilon\} \subset G.$$

Select a polygonal function y so that $\|y - x_0\| < \epsilon/3$ and $\epsilon/3 < r < \epsilon/2$, for r a rational number. Then

$$\{x(\cdot) \in C_0 \mid \|x - y\| < r\} \subset \{x(\cdot) \in C_0 \mid \|x - x_0\| < \epsilon\}.$$

The open set G in C_0 is a union of countably many open balls with centers in the countable collection of polygonal functions and rational radii.

We have shown that every open set in the metric space C_0 with sup norm is a member of $\mathcal{B}^{[0,1]}$. We write $\mathcal{B}(C_0) \subset \mathcal{B}^{[0,1]}$.

Example 9.3.4. We will show that the Borel cylinder

$$\{x(\cdot) \in C_0 \mid a_1 \leq x(t_1) < b_1, a_2 \leq x(t_2) < b_2, \ldots, a_n \leq x(t_n) < b_n\}$$

is a member of $\mathcal{B}(C_0)$.

Given $\cap_{k=1}^{n} \{x(\cdot) \in C_0 \mid a_k \leq x(t_k) < b_k\}$, we will define a projection P_{t_k} on C_0 by $P_{t_k}(x) = x(t_k)$. This projection P_{t_k} is a continuous function from C_0 to $\mathbf{R}^1$. That is, for $\|x - y\| < \delta$, we have

$$\left|P_{t_k}(x) - P_{t_k}(y)\right| = |x(t_k) - y(t_k)| \leq \|x - y\| < \delta.$$

Thus

$$\{x(\cdot) \in C_0 \mid a_k \leq x(t_k) < b_k\} = \{x(\cdot) \in C_0 \mid P_{t_k}^{-1}([a_k, b_k))\}$$

is a Borel set in C_0. We conclude that $\mathcal{B}^{[0,1]} \subset \mathcal{B}(C_0)$. Since we have already shown that $\mathcal{B}(C_0) \subset \mathcal{B}^{[0,1]}$, we may conclude that $\mathcal{B}^{[0,1]} = \mathcal{B}(C_0)$.

We have achieved an understanding of Wiener measurable sets. The next step is measurable functions.

9.4   Measurable Functionals

Definition 9.4.1 (Wiener Measurable Functional). A real-valued function F defined on C_0 is a Wiener measurable functional if it is measurable with respect to $\mathcal{B}^{[0,1]}$. That is, $F^{-1}((c, \infty))$ must be a member of $\mathcal{B}^{[0,1]}$ for all real numbers c.

Example 9.4.1. For $x \in C_0$ and $0 < t_1 \leq 1$, define $F[x] = x(t_1)$. We will show that F is a Wiener measurable functional on C_0.

Because F assigns to every function in C_0 its value at t_1, we have that

$$F^{-1}((c, \infty)) = \{x(\cdot) \in C_0 \mid F[x] \in (c, \infty)\}$$
$$= \{x(\cdot) \in C_0 \mid x(t_1) \in (c, \infty)\}.$$

This is a Borel cylinder with one restriction, a member of $\mathcal{B}^{[0,1]}$.

Example 9.4.2. For $x \in C_0$ and $0 < t_1 \leq 1$, define $F[x] = |x(t_1)|$. Is F a Wiener measurable functional on C_0?

We have two possibilities. If $c < 0$, then

$$F^{-1}((c, \infty)) = \{x(\cdot) \in C_0 \mid c < |x(t_1)| < \infty\} = C_0.$$

If $c \geq 0$, then

$$F^{-1}((c, \infty)) = \{x(\cdot) \in C_0 \mid c < |x(t_1)| < \infty\}$$
$$= \{x(\cdot) \in C_0 \mid -\infty < x(t_1) < -c\}$$
$$\cup \{x(\cdot) \in C_0 \mid c < x(t_1) < \infty\}.$$

We have Borel cylinders, members of $\mathcal{B}^{[0,1]}$. So F is a Wiener measurable functional on C_0.

Exercise 9.4.1. For $x \in C_0$, suppose $F[x] = x^2(t_1)$, where $0 < t_1 \leq 1$. Show that F is a Wiener measurable functional on C_0.

How far can we take this approach? Consider $F[x] = x^2(t_1)|x(t_2)|$, with $x \in C_0$ and $0 < t_1 < t_2 \leq 1$. We have

$$F^{-1}((c, \infty)) = \{x(\cdot) \in C_0 \mid c < x^2(t_1)|x(t_2)| < \infty\}.$$

If $c = 0$, then $x^2(t_2)|x(t_2)| > 0$ unless $x(t_1) = 0$, $x(t_2) \neq 0$, $x(t_1) \neq 0$, $x_2(t) = 0$, $x(t_1) = x(t_2) = 0$. Thus, ...

There must be a better way.

Example 9.4.3. Suppose the functional F defined on C_0 depends on only a finite number of fixed values of t. Assume $F[x] = f(x(t_1), \ldots, x(t_n))$, and f is a real-valued continuous function on $\mathbb{R}^n$. We will show that F is then a Wiener measurable functional on C_0:

$$F^{-1}((c, \infty)) = \{x(\cdot) \in C_0 \mid F[x] > c\}$$
$$= \{x(\cdot) \in C_0 \mid f(x(t_1), \ldots, x(t_n)) > c\}$$
$$= \{x(\cdot) \in C_0 \mid (x(t_1), \ldots, x(t_n)) \in f^{-1}((c, \infty))\}$$

Because f is continuous on R^n, then $f^{-1}\big((c,\infty)\big)$ is an open set, a Borel set, in R^n, say B^n. Then

$$F^{-1}\big((c,\infty)\big) = \big\{x(\cdot) \in C_0 \mid \big(x(t_1),\ldots,x(t_n)\big) \in B^n\big\},$$

a member of $\mathcal{B}^{[0,1]}$. So F is a Wiener measurable functional on C_0.

We have explored a sigma algebra of Wiener measurable subsets of C_0, $\mathcal{B}^{[0,1]}$, the Borel sets C_0 with sup norm, and Wiener measurable functionals on C_0, $F[x]$. Turning now to the Wiener integral itself, we shall mimic the development of the Lebesgue integral.

9.5   The Wiener Integral

With a measure space, $(C_0, \mathcal{B}^{[0,1]}, \mu_w)$ and measurable functions — Wiener measurable functionals on C_0, $F[x]$ — only the integral remains. Given an element $x(\cdot)$ in C_0, we can assign a real number $F[x]$. In fact, we can assign $F[x]$ to each path $x(\cdot)$ in C_0.

Thinking of this *path integral* as an averaging process, we will assign the number $F[x]$ to a collection of paths that are "close" (so to speak). Then we average over all such collections. We are now in position to define the Wiener integral. Symbolically, we write

$$\int_{C_0} F[x]\,d\mu_w.$$

As with the Lebesgue integral, we will define the Wiener integral for characteristic functionals, simple functionals, limits of simple functionals, and so on.

Example 9.5.1. Suppose $F[x]$ is the characteristic functional on a measurable subset C of C_0, where $C \in \mathcal{B}^{[0,1]}$. We will define the Wiener integral of the characteristic functional $F[x]$.

First, we observe that if $x(\cdot) \in C$, then $F[x] = 1$ and if $x \notin C$, then $F[x] = 0$. Note too that $F[x]$ is a Wiener measurable functional on C_0:

$$F^{-1}\big((-\infty,c]\big) = \begin{cases} \phi & \text{if } c < 0, \\ C_0 - C & \text{if } 0 \leq c < 1, \\ C_0 & \text{if } 1 \leq c. \end{cases}$$

Then $\int_{C_0} F[x]\,d\mu_w$ *should be* $\mu_w(C)$. We define it as such:

$$\int_{C_0} \chi_C[x]\,d\mu_w \equiv \mu_w(C).$$

Wiener simple functionals are linear combinations of characteristic functionals by definition.

Example 9.5.2. Suppose $\phi = \sum_1^n \alpha_k \chi_{C_k}$, with $C_0 = \cup_1^n C_k$, $C_i \cap C_j = \phi$, C_k members of $\mathcal{B}^{[0,1]}$.

We define the Wiener integral of ϕ by

$$\int_{C_0} \phi[x]\, d\mu_w \equiv \sum_1^n \alpha_k \mu_w(C_k).$$

How do we deal with more complicated functionals?

Example 9.5.3. Assume $F[x] = x^2(t_1)$, with $0 < t_1 \leq 1$, as suggested by Figure 9. The functional F weights each path in C_0 by the square of its value at at t_1. We shall determine

$$\int_{C_0} F[x]\, d\mu_w = \int_{C_0} x^2(t_1)\, d\mu_w.$$

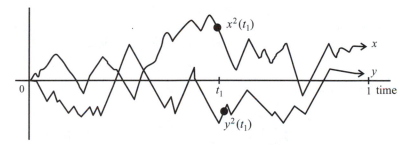

Figure 9. Wiener simple functional

Following the ideas of Lebesgue, we will construct a monotone sequence of Wiener simple functionals converging to $F[x]$. Recall that $F[x]$ is Wiener measurable (Exercise 9.4.1).

Step 1. Consider the case $[0, \infty) = [0, \frac{1}{2}) \cup [\frac{1}{2}, 1) \cup [1, \infty)$. We have

$$C_{11} = \left\{ x(\cdot) \in C_0 \,\middle|\, 0 \leq x^2(t_1) < \frac{1}{2} \right\},$$

$$C_{12} = \left\{ x(\cdot) \in C_0 \,\middle|\, \frac{1}{2} \leq x^2(t_1) < 1 \right\}, \qquad \text{and}$$

$$C_1 = \{ x(\cdot) \in C_0 \,\middle|\, 1 \leq x^2(t_1) \}.$$

Then C_{11}, C_{12}, and C_1 are disjoint members of $\mathcal{B}^{[0,1]}$ with union C_0. Define the simple functional $F_1[x]$ by

$$F_1[x] = 0 \cdot \chi_{C_{11}}[x] + \frac{1}{2} \cdot \chi_{C_{12}}[x] + 1 \cdot \chi_{C_1}[x].$$

Repeat this process.

Step n. Consider the case $[0, \infty) = \cup_{k=0}^{n2^n-1} C_{nk} \cup C_n$ with

$$C_{nk} = \left\{ x(\cdot) \in C_0 \;\Big|\; \frac{k}{2^n} \leq x^2(t_1) < \frac{k+1}{2^n} \right\}, \qquad 0 \leq k \leq n2^n - 1, \quad \text{and}$$

$$C_n = \{ x(\cdot) \in C_0 \mid n \leq x^2(t_1) \}.$$

The simple functional on C_0 is defined by

$$F_n[x] = \sum_{k=0}^{n2^n-1} \frac{k}{2^n} \chi_{C_{nk}}[x] + n\chi_{C_n}[x].$$

The reader may show that $0 \leq F_n[x] \leq F_{n+1}[x]$ on C_0, and that for $n > x^2(t_1)$ then $0 \leq F[x] - F_n[x] \leq 1/2^n$.

The Wiener measurable functional $F[x]$ is the limit of a monotone sequence of nonnegative Wiener measurable simple functionals on C_0:

$$\int_{C_0} F_n[x]\, d\mu_w = \sum_{k=0}^{n2^n-1} \frac{k}{2^n} \mu_w(C_{nk}) + n\mu_w(C_n)$$

$$= \sum_{k=0}^{n2^n-1} \left[L\int_{\sqrt{k/2^n}}^{\sqrt{(k+1)/2^n}} \left(\frac{k}{2^n} - \xi_1^2 + \xi_1^2 \right) K(\xi_1, t_1) d\xi_1 \right.$$

$$\left. + L\int_{-\sqrt{(k+1)/2^n}}^{-\sqrt{k/n}} \left(\frac{k}{2^n} - \xi_1^2 + \xi_1^2 \right) K(\xi_1, t_1) d\xi_1 \right]$$

$$+ L\int_{\sqrt{n}}^{\infty} (n - \xi_1^2 + \xi_1^2) K(\xi_1, t_1) d\xi_1$$

$$+ L\int_{-\infty}^{-\sqrt{n}} (n - \xi_1^2 + \xi_1^2) K(\xi_1, t_1) d\xi_1$$

$$= L\int_{-\infty}^{\infty} \xi_1^2 K(\xi_1, t_1) d\xi_1$$

$$+ \sum_{k=0}^{n2^n-1} \left[L\int_{\sqrt{k/2^n}}^{\sqrt{(k+1)/2^n}} \left(\frac{k}{2^n} - \xi_1^2 \right) K(\xi_1, t_1) d\xi_1 \right.$$

$$+ L \int_{-\sqrt{(k+1)/2^n}}^{-\sqrt{k/2^n}} \left(\frac{k}{2^n} - \xi_1^2 \right) K(\xi_1, t_1) d\xi_1 \Bigg]$$

$$+ L \int_{\sqrt{n}}^{\infty} (n - \xi_1^2) K(\xi_1, t_1) d\xi_1$$

$$+ L \int_{-\infty}^{-\sqrt{n}} (n - \xi_1^2) K(\xi_1, t_1) d\xi_1$$

$$\leq L \int_{-\infty}^{\infty} \xi_1^2 K(\xi_1, t_1) d\xi_1.$$

The reader may show that the four integrals with square roots as limits converge to zero as n becomes large. For example, $k/2^n - \xi_1^2 < 1/2^n$ on the interval $\left(\sqrt{k/2^n}, \sqrt{(k+1)/2^n} \right)$ and the sum $\sum_{k=0}^{n2^n-1} [\ldots]$ is dominated by $1/2^n \int_{-\infty}^{\infty} K(\xi_1, t_1) d\xi$. As for $L \int_{\sqrt{n}}^{\infty} (n - \xi_1^2) K(\xi_1, t_1) d\xi_1$, use $\int_{\alpha}^{\infty} e^{-x^2} dx < 1/2\alpha e^{-\alpha^2}$.

Thus,

$$\lim \int_{C_0} F_n[x] d\mu_w = L \int_{-\infty}^{\infty} \xi_1^2 K(\xi_1, t_1) d\xi_1 = t_1.$$

We define

$$\int_{C_0} F[x] d\mu_w \equiv \lim \int_{C_0} F_n[x] d\mu_w.$$

In this example, $F[x] = x^2(t_1)$, the Wiener integral is evaluated as an elementary Lebesgue integral.

Example 9.5.4. Given $F[x] = x^2(t_1)|x(t_2)|$, calculate

$$\int_{C_0} x^2(t_1)|x(t_2)| d\mu_w, \qquad \text{for } 0 < t_1 < t_2 \leq 1.$$

First we partition R^1:

$$(-\infty, -n) \cup \left[-n, -n + \frac{1}{2^n} \right) \cup \cdots \cup \left[n - \frac{1}{2^n}, n \right) \cup [n, \infty).$$

Let

$$C_n^- = \{x(\cdot) \in C_0 \mid -\infty < x(t_1) < -n\},$$

$$C_{nk} = \left\{ x(\cdot) \in C_0 \mid \frac{k}{2^n} \leq x(t_1) < \frac{k+1}{2^n} \right\},$$

$$C_n^+ = \{x(\cdot) \in C_0 \mid n \leq x(t_1)\}.$$

We have partitioned C_0 into a finite collection of disjoint Borel cylinders, and if we replace $x(t_1)$ with ξ_1 we have a partition of R^1 into a finite collection of disjoint Borel sets in R^1: B_n^-, B_{nk}, $and B_n^+$, respectively.

Repeat the process for $x(t_2)$, letting

$$D_n^- = \{x(\cdot) \in C_0 \mid -\infty < x(t_2) < -n\},$$
$$D_{ni} = \left\{x(\cdot) \in C_0 \mid \frac{i}{2^n} \le x(t_2) < \frac{i+1}{2^n}\right\},$$
$$D_n^+ = \{x(\cdot) \in C_0 \mid n \le x(t_2)\};$$

with corresponding disjoint Borel sets in R^1: G_n^-, G_{ni}, G_n^+.

Together, we have a partitioning of $R^2 = R^1 \times R^1$ into disjoint Borel sets that correspond to disjoint Borel cylinders in $\mathcal{B}^{[0,1]}$. The diagram in Figure 10 indicates the appropriate regions.

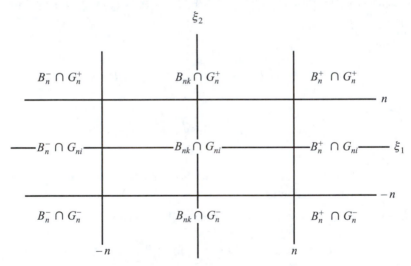

Figure 10. Partition of $R^1 \times R^1$

Adding (vertically) these approximations and taking limits we obtain

$$L \int_{-\infty}^{\infty} d\xi_2 \int_{-\infty}^{n} d\xi_1 \xi_1^2 |\xi_2| K_1 K_2 + L \int_{-\infty}^{\infty} d\xi_2 \int_{-n}^{n} d\xi_1 \xi_1^2 |\xi_2| K_1 K_2$$
$$+ L \int_{-\infty}^{\infty} d\xi_2 \int_{n}^{\infty} d\xi_1 \xi_1^2 |\xi_2| K_1 K_2,$$

for $K_1 K_2 = K(\xi_1, t_1) K(\xi_2 - \xi_1, t_2 - t_1)$, yielding

$$L \int_{-\infty}^{\infty} d\xi_2 \int_{-\infty}^{\infty} d\xi_1 \xi_1^2 |\xi_2| K(\xi_1, t_1) K(\xi_2 - \xi_1, t_2 - t_1).$$

Routine (though lengthy) calculations will yield

$$\lim \int_{C_0} F_n[x] \, d\mu_w = \mathrm{L} \int_{-\infty}^{\infty} \int_{-\infty}^{\infty} \xi_1^2 |\xi_2| K(\xi_1, t_1) K(\xi_2 - \xi_1, t_2 - t_1) \, d\xi_1 d\xi_2.$$

From these last two examples,

$$\int_{C_0} x^2(t_1) \, d\mu_w = \mathrm{L} \int_{-\infty}^{\infty} \xi_1^2 K(\xi_1, t_1) \, d\xi_1$$

and

$$\int_{C_0} x^2(t_1) \, |x(t_2)| \, d\mu_w$$

$$= \mathrm{L} \int_{-\infty}^{\infty} \int_{-\infty}^{\infty} \xi_1^2 |\xi_2| K(\xi_1, t_1) K(\xi_2 - \xi_1, t_2 - t_1) \, d\xi_1 d\xi_2.$$

The Wiener integrals have reduced to Lebesgue integrals of the same form.

9.6 Functionals Dependent on a Finite Number of t Values

Theorem 9.6.1. *Suppose a functional $F[x]$ defined on C_0 depends on a finite number of fixed t values, $0 < t_1 < t_2 < \cdots < t_n \leq 1$ such that $F[x] = f\big(x(t_1), x(t_2), \ldots, x(t_n)\big)$, where f is a real-valued function continuous on R^n.*

Then F is a Wiener measurable functional on C_0, and

$$\int_{C_0} F[x] \, d\mu_w = \int_{C_0} f\big(x(t_1), \ldots, x(t_n)\big) \, d\mu_w$$

$$= \mathrm{L} \int_{-\infty}^{\infty} d\xi_n \cdots \int_{-\infty}^{\infty} d\xi_1 f(\xi_1, \ldots, \xi_n) K(\xi_1, t_1)$$

$$\cdots K(\xi_n - \xi_{n-1}, t_n - t_{n-1})$$

whenever the last integral exists.

If the Wiener functional depends continuously on only a finite number of t values, it may be evaluated as an ordinary Lebesgue integral.

Proof. We will sketch the proof. As discussed in Example 9.4.3, $F[x]$ is a Wiener measurable functional on C_0:

$$F^{-1}\big((c, \infty)\big) = \left\{ x(\cdot) \in C_0 \mid \big(x(t_1), \ldots, x(t_n)\big) \in \underset{\text{Borel}}{f^{-1}(c, \infty) \subseteq \mathrm{R}^n} \right\}.$$

We can mimic the development of the Lebesgue integral. Assume f is nonnegative and continuous. This is true for characteristic functionals, simple functionals, and limits of monotone sequences $\{F_m\}$ of simple functionals.

(The reader may partition R^n into nonoverlapping Borel rectangles. On each of the associated Borel cylinders, calculate the infimum of f times the Wiener measure of the associated Borel cylinder.)

To conclude, we calculate

$$\int_{C_0} F[x] \, d\mu_w = \int_{C_0} f\big(x(t_1), \dots, x(t_n)\big) \, d\mu_w = \lim_m \int_{C_0} F_m[x] \, d\mu_w$$

$$= \int \cdots \int_{R^n} f(\xi_1, \xi_2, \dots, \xi_n) K(\xi_1, t_1)$$

$$\cdots K(\xi_n - \xi_{n-1}, t_n - t_{n-1}) \, d\xi_1 \, d\xi_2 \dots d\xi_n,$$

by monotone convergence. For f continuous, proceed as $(f + |f|)/2$, $(|f| - f)/2$, and so on.

Exercise 9.6.1. Calculate $\int_{C_0} F[x(\cdot)] \, d\mu_w$ for the following functionals F.

a. $F[x(\cdot)] = x(t_1)$, $0 < t_1 \leq 1$.

b. $F[x(\cdot)] = x^2(t_1)$, $0 < t_1 \leq 1$.

c. $F[x(\cdot)] = x^2(t_1) \, |x(t_2)|$, $0 < t_1 < t_2 \leq 1$.

d. Show

$$\int_{C_0} x^n(t_1) \, d\mu_w = \begin{cases} 0 & n \text{ odd}, \\ 1 \cdot 3 \cdots \cdots (n-1) t_1^{n/2} & n \text{ even}. \end{cases}$$

e. Show $\int_{C_0} x(t_1) x(t_2) \, d\mu_w = t_1$, $0 < t_1 < t_2 \leq 1$.

f. $F[x(\cdot)] = x(t_2) - x(t_1)$, $0 < t_1 < t_2 \leq 1$. Hint: $\mu_w\big(\{x \in C_0 | x(t_2) - x(t_1) \leq \beta\}\big) = \mu_w\big(\{x \in C_0 | (x(t_1), x(t_2)) : (\xi_1, \xi_2) \in B^2 \text{ with } \xi_2 - \xi_1 \leq \beta\}\big) = L \int B^2 \int (\xi_2 - \xi_1) K(\xi_1, t_1) K(\xi_2 - \xi_1, t_2 - t_1) \, d(\xi_1, \xi_2)$. Let $z_1 = \xi_1$, $z_2 = \xi_2 - \xi_1$, and use Fubini's Theorem.

g. $F[x(\cdot)] = \big(x(t_2) - x(t_1)\big)^2$, $0 < t_1 < t_2 \leq 1$.

h. Show $\int_{C_0} \big(x(t_2) - x(t_1)\big)^4 \, d\mu_w = 3(t_2 - t_1)^2$, $0 < t_1 < t_2 \leq 1$.

i. $F[x(\cdot)] = \big(x(t_2) - x(t_1)\big)^n$, $0 < t_1 < t_2 \leq 1$.

j. Show $\int_{C_0} \big(x(t_2) - x(t_1)\big)\big(x(t_4) - x(t_3)\big) \, d\mu_w = 0$, $0 < t_1 < t_2 \leq t_3 < t_4 \leq 1$.

9.6.1   More Interesting Functionals

What if F depends on more than a finite number of values of t?

Example 9.6.1. Suppose $F[x] = C \int_0^1 |x(\tau)| \, d\tau$. The continuous function $x(\cdot)$, the path $x(\cdot)$, is to be weighted by the integral of its absolute value.

Let

$$F_N[x] = \frac{1}{N} \sum_{k=1}^{N} \left| x\left(\frac{k}{N}\right) \right|,$$

an approximation to $C \int_0^1 |x(\tau)| \, d\tau$. (Realize that

$$f(\xi_1, \xi_2, \ldots, \xi_N) \equiv \frac{|\xi_1| + |\xi_2| + \cdots + |\xi_N|}{N}$$

is a continuous function of R^N.)

By Example 9.4.3, $F_N[x]$ is a Wiener measurable functional on C_0, and by Theorem 9.6.1,

$$\int_{C_0} F_N[x] \, d\mu_w = \frac{1}{N} \int_{-\infty}^{\infty} d\xi_N \cdots \int_{-\infty}^{\infty} d\xi_1 (|\xi_1| + \cdots + |\xi_N|) K(\xi_1, t_1)$$
$$\cdots K(\xi_N - \xi_{N-1}, t_N - t_{N-1}).$$

We want $\int_{C_0} F[x] \, d\mu_w = \lim_N \int_{C_0} F_N[x] \, d\mu_w$. We have

$$\left| \int_0^1 |x(\tau)| \, d\tau - \frac{1}{N} \sum_{k=1}^{N} \left| x\left(\frac{k}{N}\right) \right| \right|$$

$$= \left| \sum_{k=1}^{N} \int_{(k-1)/N}^{k/N} \left(|x(\tau)| - \left| x\left(\frac{k}{N}\right) \right| \right) d\tau \right|$$

$$\leq \sum_{k=1}^{N} \int_{(k-1)/N}^{k/N} \left| |x(\tau)| - \left| x\left(\frac{k}{N}\right) \right| \right| d\tau$$

$$\leq \sum_{k=1}^{n} \int_{(k-1)/N}^{k/N} \left| x(\tau) - x\left(\frac{k}{N}\right) \right| d\tau.$$

Because $x(\cdot)$ is uniformly continuous on $[0, 1]$, this last sum may be made arbitrarily small. That is, $\lim_N F_N[x] = F[x]$.

The function $F[x]$ is a nonnegative, Wiener measurable functional on C_0 as a limit of Wiener measurable functions F_N on C_0. As for evaluating the

Wiener integral of F, using Fubini's Theorem 6.7.1, we have

$$\int_{C_0} \left(\text{C} \int_0^1 |x(\tau)| \, d\tau \right) d\mu_w = \text{L} \int_0^1 \left(\int_{C_0} |x(\tau)| \, d\mu_w \right) d\tau$$

$$= \text{L} \int_0^1 \left(\int_{-\infty}^{\infty} |\xi| \, K(\xi, \tau) d\xi \right) d\tau$$

$$= \text{L} \int_0^1 \left(\sqrt{\frac{2}{\pi}} \tau^{1/2} \right) d\tau = \sqrt{\frac{2}{\pi}} \left(\frac{2}{3} \right).$$

Evaluating as a limit, we have

$$\int_{C_0} F_N[x] \, d\mu_w = \sum_{k=1}^N \frac{1}{N} \int_{C_0} \left| x \left(\frac{k}{N} \right) \right| \, d\mu_w$$

$$= \sum_{k=1}^N \frac{1}{N} \text{L} \int_{-\infty}^{\infty} |\xi_k| \, K(\xi_k, \tau) d\xi_k$$

$$= \sqrt{\frac{2}{\pi}} \sum_{k=1}^N \sqrt{\frac{k}{N}} \cdot \frac{1}{N}$$

and

$$\lim_N \int_{C_0} F_N[x] \, dx = \lim_N \sqrt{\frac{2}{\pi}} \left(\sum_{k=1}^N \sqrt{\frac{k}{N}} \cdot \frac{1}{N} \right)$$

$$= \sqrt{\frac{2}{\pi}} \int_0^1 \tau^{1/2} d\tau = \sqrt{\frac{2}{\pi}} \left(\frac{2}{3} \right).$$

Exercise 9.6.2. Calculate $\int_{C_0} F[x] \, d\mu_w$ for the following cases ($0 < t \leq 1$).

   a. $F[x] = \text{L} \int_0^t |x(\tau)|^m \, d\tau, m = 1, 2, \ldots$.

   b. $F[x] = \text{L} \int_0^t x^{2m-1}(\tau) d\tau, m = 1, 2, \ldots$.

   c. $F[x] = \text{L} \int_0^t x^{2m}(\tau) d\tau, m = 1, 2, \ldots$.

Exercise 9.6.3. Suppose

$$F[x] = \exp \left(-\text{L} \int_0^t x^2(\tau) d\tau \right) \quad \text{and}$$

$$F_N[x] = \exp \left(-\frac{t}{N} \sum_{k=1}^N x^2 \left(k \frac{t}{N} \right) \right).$$

a. Argue Wiener measurability on C_0 of $F_N[x]$ and $F[x]$.

b. Because $|F_N[x]| \leq 1$, use the Bounded Convergence Theorem 6.3.1 to conclude that

$$\int_{C_0} \exp\left(-L \int_0^t x^2(\tau)d\tau\right) d\mu_w$$

$$= \lim_N L \int_{-\infty}^{\infty} \cdots \int_{-\infty}^{\infty} \exp\left(-\frac{t}{N} \sum_{k=1}^n \xi_k^2\right)$$

$$\times K(\xi_1, t_1) \cdots K(\xi_n - \xi_{n-1}, t_n - t_{n-1}) d\xi_1 \ldots d\xi_n,$$

with $t_k = kt/N$.

Exercise 9.6.4. Given $F[x] = \exp\left(-L \int_0^t V\left(x(\tau)\right)d\tau\right)$, where $0 \leq V(\xi) \leq M \,\forall \xi \in R^1$, with V continuous on R^1. Proceeding as above,

$$F_N[x] = \exp\left(-\frac{t}{N} \sum_{k=1}^N V\left(k\frac{t}{N}\right)\right).$$

Show that

$$\int_{C_0} F[x] \, d\mu_w = \lim_N \left(2\pi \frac{t}{N}\right)^{-N/2} \cdot L \int_{-\infty}^{\infty} \cdots \int_{-\infty}^{\infty}$$

$$\exp\left(-\frac{t}{N} \sum_{k=1}^N \left[V(\xi_k) + \left(\frac{\xi_k - \xi_{k-1}}{t/N}\right)^2\right]\right) d\xi_1 \ldots d\xi_n,$$

where $\xi_0 = 0$.

Note that

$$\frac{t}{N} \sum_{k=1}^N \left[V(\xi_k) + \left(\frac{\xi_k - \xi_{k-1}}{t/N}\right)^2\right]$$

is a discrete approximation of

$$\int_0^t \left[V\left(x(\tau)\right) + \left(\frac{dx}{d\tau}\right)^2\right] d\tau.$$

But $V(x) + (dx/dt)^2$ is the Hamiltonian of a particle of mass 1 in a field V. The Feynman integral (next chapter) involves the Lagrangian.

9.7   Kac's Theorem

As the reader has observed, the Wiener integrals are often the limits of fairly obvious discrete integrals, just as Riemann integrals often may be evaluated as limits of discrete sums. As with the Riemann integral, this arduous process sometimes may be avoided.

We have a remarkable theorem due to Mark Kac that relates the evaluation of a class of Wiener integrals to solutions to appropriate integral equations. Essentially, he shows that a very complicated Wiener integral is related to a solution of an integral equation.

The proof of this theorem may be found many places, including Kac (1959). As a tribute to his genius, we conclude our treatment of the Wiener integral with Kac's proof of this theorem.

Theorem 9.7.1 (Kac, 1949). *Suppose that $V(\xi)$ is a continuous nonnegative bounded function on $\mathbf{R}^1$. Then*

$$\int_{C_0} \exp\left(-\int_0^t V(x(\tau))d\tau\right) d\mu_w = L \int_{-\infty}^{\infty} H(x,t))\,dx, \qquad 0 < t \le 1,$$

where H satisfies the integral equation

$$H(x,t) + L \int_0^t d\tau \int_{-\infty}^{\infty} d\xi K(x-\xi, t-\tau)V(\xi)K(\xi,\tau) = K(x,t).$$

Recalling Exercise 9.1.5, note that for $V \equiv 0$ we have

$$1 = \int_{C_0} 1\,d\mu_w = \int_{-\infty}^{\infty} K(x,t)\,dx, \qquad \text{where } K_t = \frac{1}{2}K_{xx}.$$

Proof. Assume $0 \le V(\xi) \le M$ for all ξ in $\mathbf{R}^1$. There are five stages to this proof.

Step 1. By Exercise 9.6.4, Wiener measurability of $\exp\left(-L \int_0^t V(x(\tau))d\tau\right)$ is assured. Furthermore, for $0 \le L \int_0^t V(x(\tau))d\tau \le Mt$ we have

$$\exp\left(-L \int_0^t V(x(\tau))d\tau\right) = \sum_{k=0}^{\infty}(-1)^k \frac{\left[L \int_0^t V(x(\tau))d\tau\right]^k}{k!}$$

and application of the Bounded Convergence Theorem implies that

$$\int_{C_0} \exp\left(-L\int_0^t V(x(\tau))d\tau\right) d\mu_w$$

$$= \sum_{k=0}^{\infty} \frac{(-1)^k}{k!} \int_{C_0} \left(L\int_0^t V(x(\tau))d\tau\right)^k d\mu_w.$$

Step 2. Let $H_0(x,t) = K(x,t) = (2\pi t)^{-1/2} \exp\left(-x^2/2t\right)$ and define

$$H_k(x,t) = L\int_0^t d\tau \int_{-\infty}^{\infty} d\xi K(x-\xi, t-\tau)V(\xi)H_{k-1}(\xi,\tau).$$

Show

$$\int_{-\infty}^{\infty} H_k(x,t)\, dx = \frac{1}{k!}\int_{C_0}\left(L\int_0^t V(x(\tau))d\tau\right)^k d\mu_w, \qquad k = 1, 2, \ldots$$

and

$$L\int_{-\infty}^{\infty} H_0(x,t)\, dx = L\int_{-\infty}^{\infty} K(x,t)\, dx = 1, \qquad t > 0.$$

Hint: $\left(L\int_0^t V(x(\tau))d\tau\right)^2 = 2!L\int_0^t d\tau_2 \int_0^{\tau_2} d\tau_1 V(x(\tau_1))V(x(\tau_2)).$

Step 3. Estimating $H_k(x,t)$, we have

$$0 \le H_1(x,t) = L\int_0^t d\tau \int_{-\infty}^{\infty} d\xi K(x-\xi, t-\tau)V(\xi)H_0(\xi,\tau)$$

$$\le MtH_0(x,t).$$

$0 \le H_2(x,t)$

$$\le M^2 L \int_0^t d\tau_2 \int_0^{\tau_2} d\tau_1 \int_{-\infty}^{\infty} d\xi_2 \int_{-\infty}^{\infty} d\xi_1$$

$$\times K(x-\xi_2, t-\tau_2)K(\xi_2-\xi_1, \tau_2-\tau_1)K(\xi_1, \tau_1)$$

$$= M^2 L \int_0^t d\tau_2 \int_0^{\tau_2} d\tau_1 \int_{-\infty}^{\infty} d\xi_2 K(x-\xi_2, t-\tau_2)K(\xi_2, \tau_2) \quad \text{(by 9.2.2)}$$

$$= M^2 L \int_0^t d\tau_2 \int_0^{\tau_2} d\tau_1 K(x,t) = \frac{(Mt)^2}{2!} H_0(x,t).$$

In general,

$$0 \le H_k(x,t) \le \frac{(Mt)^k}{k!} H_0(x,t), \qquad k = 1, 2, \ldots.$$

Step 4. Define $H(x,t) = \sum_{k=0}^{\infty} (-1)^k H_k(x,t)$, which converges for all x and $t > 0$, since we have $|H(x,t)| \le e^{Mt} H_0(x,t)$.

Step 5. Then

$$
\begin{aligned}
H(x,t) &= H_0(x,t) + \sum_{k=1}^{\infty} (-1)^k H_k(x,t) \\
&= H_0(x,t) + \sum_{k=1}^{\infty} (-1)^k \, \mathrm{L} \int_0^t d\tau \\
&\qquad \int_{-\infty}^{\infty} d\xi K(x-\xi, t-\tau) V(\xi) H_{k-1}(\xi, \tau) \\
&= H_0(x,t) - \mathrm{L} \int_0^t d\tau \\
&\qquad \int_{-\infty}^{\infty} d\xi K(x-\xi, t-\tau) V(\xi) \left(\sum_{k=0}^{\infty} (-1)^k H_k(\xi, \tau) \right) \\
&= H_0(x,t) - \mathrm{L} \int_0^t d\tau \int_{-\infty}^{\infty} d\xi K(x-\xi, t-\tau) V(\xi) H(\xi, \tau).
\end{aligned}
$$

That is,

$$
H(x,t) + \mathrm{L} \int_0^t d\tau \int_{-\infty}^{\infty} d\xi K(x-\xi, t-\tau) V(\xi) H(\xi, \tau) = K(x,t), \qquad t > 0,
$$

and

$$
\int_{C_0} \left(\exp\left(-\mathrm{L} \int_0^t V\big(x(\tau)\big) d\tau \right) \right) d\mu_w = \mathrm{L} \int_{-\infty}^{\infty} H(x,t)\, dx
$$

for all x with $t > 0$. $\square$

The requirement that V be bounded on R^1 may be removed by truncation arguments.

9.8   References

1. Kac, Mark *Probability and Related Topics in Physical Sciences.* London: Interscience Publishers, 1959.

2. Wiener, Norbert. *Collected Works.* Vol. I. Cambridge, Mass.: MIT Press, 1976.

3. Yeh, J. *Stochastic Processes and the Wiener Integral.* New York: Marcel Dekker, 1973.

CHAPTER **10**

The Feynman Integral

[Mathematics]...there is no study in the world which brings into more harmonious action all the faculties of the mind. — J. J. Sylvester

10.1 Introduction

In the 1920s Norbert Wiener introduced the concept of *a measure on the space of continuous functions*. As you recall from Chapter 9, this idea arose from his attempts to model the Brownian motion of small particles suspended in a fluid. In the 1940s Richard Feynman (1918–1988) developed an integral on the same space of continuous functions in his efforts to model the quantum mechanics of very small particles such as electrons. To succeed, Feynman's *path integral* approach to quantum mechanics had to be consistent with Schrödinger's Equation.

10.1.1 Schrödinger's Equation

A frequent correspondent of Albert Einstein, Erwin Schrödinger (1887–1961) made a series of brilliant advances in quantum theory and the general theory of relativity. Our topic here is his breakthrough wave equation, discovered in 1925.

Suppose a particle of mass m is at position x_0 at time $t = 0$ with a potential $V(x_0)$. The particle may move to position x at time t. The Heisenberg Uncertainty Principle sets accuracy limits on the determination of position x at time t. From physical considerations, then, we assign a probability to each path from x_0 at time $t = 0$ to x at time t. This probability is $P(t, x) = |\psi(t, x)|^2$, where ψ is a complex-valued quantity called the *probability amplitude*.

Schrödinger's Equation, which is the partial differential equation

$$\frac{\partial \psi}{\partial t} = \frac{i\hbar}{2m} \frac{\partial^2 \psi}{\partial x^2} - \frac{i}{\hbar} V\psi,$$

with $-\infty < x < \infty$, $t > 0$, and $\hbar = 1.054 \times 10^{-27}$ erg-sec, is satisfied by the probability amplitude ψ with $\psi(0, x) = f(x)$. Because $P(t, x) = |\psi(t, x)|^2$ is a probability, we want

$$\int_{\mathrm{R}} |\psi(t, x)|^2 \, dx = 1, \qquad \text{for } t \geq 0.$$

10.1.2 Feynman's Riemann Sums

Feynman's path integral approach exploits the idea of Riemann sums. Suppose we have a nonnegative continuous function f on the interval $[a, b]$ and we are faced with the task of determining the area of the region between the x-axis and the graph of f for x between a and b. Roughly, we can say the area is the sum of all the ordinates—the sum of all the fs. In practice, we take a finite subset of the ordinates, equally spaced, and calculate the sum.

Take another set of ordinates, equally spaced but closer together, and form another sum. Generally, as we take more and more ordinates (equally spaced, closer and closer together) and compute their sum, these sums will not approach a limit. Clearly, $\sum_{a \leq x \leq b} f(x)$ doesn't make sense.

But what if we assign a weight to each ordinate before summing? We can assign an *appropriate weight h*, a normalizing constant reflecting the spacing between consecutive ordinates. Now we have $\sum f(x_k) \cdot h$, and such sums have a limit, the so-called *Riemann integral of the function f over the interval* $[a, b]$. We write

$$hf(x_1) + hf(x_2) + \cdots + hf(x_n) \to \mathrm{R} \int_a^b f(x) \, dx.$$

It is this appropriate weight, this normalizing constant, that we will be trying to determine for each possible path.

To each possible path, Feynman postulated a probability amplitude, the squared magnitude of which was to be the probability for that particular path. All paths contribute, and each path contributes an equal amount to the total amplitude, but at different phases.

The phase of the contribution from a particular path is proportional (the normalizing constant) to the action along that path. The action for a particular path is the action for the corresponding classical system.

What does all this mean? Let's look at an example.

10.2   Summing Probability Amplitudes

Suppose a small particle of mass m is at location $x(0) = x_0$ at time 0. It moves to location $x(t) = x$ at time t in the presence of a potential $V(x)$, along a path $x(\tau)$, where $0 \leq \tau \leq t$. See Figure 1.

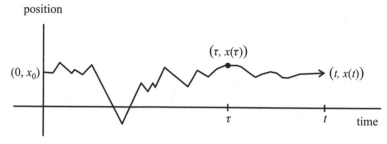

position

$(\tau, x(\tau))$

$(0, x_0)$ $(t, x(t))$

τ t time

Figure 1. Path $(\tau, x(\tau))$

We begin by isolating several key pieces of this puzzle:

- The *Lagrangian* is the difference between the kinetic energy and the potential energy, or $\frac{1}{2}m\dot{x}^2 - V(x)$.

- The *action A* along this path $x(\tau)$ is a functional given by the integral of the Lagrangian along this path: $A[x(\cdot)] = \int_0^t \left[\frac{1}{2}m\dot{x}^2(\tau) - V\left(x(\tau)\right)\right] d\tau$.

- The *phase* of the contribution from a particular path is proportional to the action along that path $e^{(i/\hbar)A[x(\cdot)]}$, where $\hbar = 1.054 \times 10^{-27}$.

- The *probability amplitude* for this path is proportional to its phase, $Ke^{(i/\hbar)A[x(\cdot)]}$, where K is the normalizing constant, the same constant for each path.

- The *total probability amplitude* ψ is the sum of the individual probability amplitudes over all continuous paths from $(0, x_0)$ to $(t, x(t))$:

$$\psi\left((0, x_0); (t, x)\right) = \sum_{\substack{\text{all connecting} \\ \text{continuous paths}}} Ke^{(i/\hbar)A[x(\cdot)]}.$$

This function ψ is to be a solution of Schrödinger's Equation, and the probability of going from x_0 at time 0 to x at time t is $|\psi|^2$. We apply the Riemann sum analogy as suggested by Feynman.

10.2.1   First Approximation

Divide the time interval $[0, t]$ into n equal parts, $x_k = x(kt/n)$, with $0 \leq k \leq n$. Now replace continuous paths with polygonal paths, using $x_n = x(t) = x$ and $t_k = kt/n$. See Figure 2.

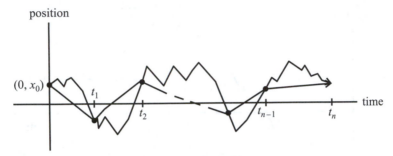

Figure 2. Polygonal approximation

10.2.2   Second Approximation

Approximate the action $A[x(\cdot)]$ along the polygonal path

$$A[x(\cdot)] = \int_0^t \left[\frac{1}{2}m\dot{x}^2(\tau) - V\left(x(\tau)\right) \right] d\tau$$

$$\approx \sum_1^n \left[\frac{1}{2}m \left(\frac{x_k - x_{k-1}}{t/n} \right)^2 - V(x_{k-1}) \right] \left(\frac{t}{n} \right)$$

$$= \sum_1^n \left[\frac{m}{2(t/n)} (x_k - x_{k-1})^2 - V(x_{k-1}) \left(\frac{t}{n} \right) \right].$$

The phase of the contribution from this polygonal path is

$$\exp \frac{i}{\hbar} \left\{ \sum_1^n \left[\frac{m}{2(t/n)} (x_k - x_{k-1})^2 - V(x_{k-1}) \left(\frac{t}{n} \right) \right] \right\}.$$

Summing over all polygonal paths, we arrive at the Feynman integral approximation of the sum of all probability amplitudes over all continuous paths,

$$\sum_{\substack{\text{all connecting} \\ \text{continuous paths}}} K e^{(i/\hbar)A[x(\cdot)]}.$$

This approximation is

$$\psi_n\big((0, x_0); (t, x)\big) = \int_{\mathrm{R}} dx_{n-1} \cdots \int_{\mathrm{R}} dx_1 K$$
$$\cdot \exp \frac{i}{\hbar} \left\{ \sum_1^n \left[\frac{m}{2(t/n)} (x_k - x_{k-1})^2 - V(x_{k-1}) \left(\frac{t}{n}\right) \right] \right\}.$$

Note: We obtain all polygonal paths as $x_1, x_2, \ldots, x_{n-1}$ vary over R; x_0 and $x_n = x$ are fixed.

Taking the limit, we have

$$\psi\big((0, x_0); (t, x)\big) = \lim_{n \to \infty} \int_{\mathrm{R}} dx_{n-1} \cdots \int_{\mathrm{R}} dx_1 K$$
$$\cdot \exp \frac{i}{\hbar} \left\{ \sum_1^n \left[\frac{m}{2(t/n)} (x_k - x_{k-1})^2 - V(x_{k-1}) \left(\frac{t}{n}\right) \right] \right\}$$

as the solution of Schrödinger's Equation,

$$\frac{\partial \psi}{\partial t} = \frac{i\hbar}{2m} \psi_{xx} - \frac{i}{\hbar} V \psi, \qquad \text{with } -\infty < x < \infty, t > 0.$$

What kind of integral is this? The integral is highly oscillatory and is not absolutely convergent when V is real.

What does the limit mean? Is it possible to choose a normalizing constant K that will guarantee a limit in some sense? Would a K that yields a limit also give us a solution to Schrödinger's Equation? In Feynman's words, "to define such a normalizing constant seems to be a very difficult problem and we do not know how to do this in general terms."

10.2.3   The Normalizing Constant

Feynman goes on to suggest that the normalizing constant K is given by

$$K = \left(\frac{m}{2\pi i \hbar(t/n)} \right)^{n/2}.$$

By employing the principle of *superposition*, that is,

$$\psi(t, x) = \int_{\mathrm{R}} \psi\big((0, x_0); (t, x)\big) f(x_0) \, dx_0, \qquad \text{with } \psi(0, x) = f(x),$$

and incorporating this normalizing constant K, we finally have Feynman's solution to what we shall call Schrödinger Problems A and B.

Schrödinger Problem A

Problem A assumes that the potential V is 0. In this case,

$$
\psi(t, x) = \lim_{n} \left(\frac{m}{2\pi i \hbar(t/n)} \right)^{n/2} \int_{\mathbb{R}} dx_{n-1} \cdots \int_{\mathbb{R}} dx_0
$$
$$
\cdot \exp\left\{ \frac{im}{2\hbar(t/n)} \sum_{1}^{n} (x_k - x_{k-1})^2 \right\} f(x_0), \qquad \text{for } x_n = x,
$$

solves

$$
\frac{\partial \psi}{\partial t} = \frac{i\hbar}{2m} \frac{\partial^2 \psi}{\partial x^2}, \qquad \text{with } -\infty < x < \infty, t > 0,
$$

$$
\psi(0, x) = f(x), \qquad \text{with } \int_{\mathbb{R}} |\psi(t, x)|^2 \, dx = 1, t \geq 0.
$$

Schrödinger Problem B

Problem B asssumes $V \neq 0$. In this case

$$
\psi(t, x) = \lim_{n} \left(\frac{m}{2\pi i \hbar(t/n)} \right)^{n/2} \int_{\mathbb{R}} dx_{n-1} \cdots \int_{\mathbb{R}} dx_0
$$
$$
\cdot \exp\left\{ \frac{im}{2\hbar(t/n)} \sum_{1}^{n} (x_k - x_{k-1})^2 - \frac{i}{\hbar}\left(\frac{t}{n}\right) \sum_{1}^{n} V(x_{k-1}) \right\} f(x_0)
$$

solves

$$
\frac{\partial \psi}{\partial t} = \frac{i\hbar}{2m} \frac{\partial^2 \psi}{\partial x^2} - \frac{i}{\hbar} V \psi \qquad \text{with } -\infty < x < \infty, t > 0,
$$

$$
\psi(0, x) = f(x), \qquad \text{with } \int_{\mathbb{R}} |\psi(t, x)|^2 \, dx = 1, t \geq 0.
$$

In the remainder of this chapter we will explain what is meant by "solves" in the context of Schrödinger Problem A and make some general comments about Problem B.

10.3   A Simple Example

It is time for an example.

Example 10.3.1. Suppose a particle of mass m at location $x(0) = x_0$ at time 0 moves to location $x(t) = x_n$ at time t in the absence of a potential field. In other words, $V = 0$. We claim that

$$\psi(t, x) = \lim_n \left(\frac{m}{2\pi i \hbar(t/n)}\right)^{n/2} \underbrace{\int_R \cdots \int_R}_{n-1}$$

$$\cdot \exp\left(\frac{im}{2\hbar(t/n)} \sum_1^n (x_k - x_{k-1})^2\right) dx_1\, dx_2 \ldots dx_{n-1},$$

where $x_n = x$, solves

$$\frac{\partial \psi}{\partial t} = \frac{i\hbar}{2m} \frac{\partial^2 \psi}{\partial x^2} \qquad \text{with } -\infty < x < \infty, t > 0.$$

Let $a = m/[2\hbar(t/n)]$. To evaluate this integral, we must integrate expressions of the form

$$\int_R \exp\left\{i\left[a(x-u)^2 + b(y-u)^2\right]\right\} du.$$

Our approach has three parts.

Part One. Show

$$\int_{-\infty}^{\infty} e^{iax^2}\, dx = \sqrt{\frac{\pi i}{a}}, \qquad \text{for } a > 0.$$

Hint: Consider the contour integral $\oint_C e^{iaz^2}\, dz$, for $a > 0$.

Write $e^{iaz^2} = u(x, y) + iv(x, y)$, and show that u, v satisfy the Cauchy–Riemann equations, $u_x = v_y$ and $u_y = -v_x$. Continuity of u, v and their partial derivatives show e^{iaz^2} is an analytic function. Consequently, with a suitable contour C to be described, $\oint_C e^{iaz^2}\, dz = 0$.

Let C be the contour for e^{iaz^2} indicated in Figure 3. Then

$$0 = \oint_C e^{iaz^2}\, dz$$

$$= \int_0^B e^{iax^2}\, dx + \int_0^B e^{ia(B+iy)^2} i\, dy + \int_B^0 e^{ia(1+i)^2 x^2}(1+i)\, dx.$$

Using $\sqrt{i} = (1+i)/\sqrt{2}$, we conclude that

$$\int_0^{\infty} e^{iax^2}\, dx = \frac{1+i}{2} \sqrt{\frac{\pi}{2a}}.$$

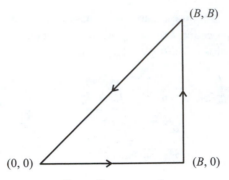

Figure 3. Contour C

Thus

$$\int_R e^{iax^2}\, dx = (1+i)\sqrt{\frac{\pi}{2a}} = \left(\frac{1}{\sqrt{2}} + \frac{i}{\sqrt{2}}\right)\sqrt{\frac{\pi}{a}} = \sqrt{\frac{i\pi}{a}}.$$

Part Two. Completing the square, show

$$\int_R du\, \exp i\left[a(x-u)^2 + b(u-y)^2\right] = \sqrt{\frac{\pi i}{a+b}}\, e^{[iab/(a+b)](x-y)^2}.$$

Conclude

$$\int_R dx_1 e^{i[a(x_1-x_0)^2 + a(x_2-x_1)^2]} = \sqrt{\frac{\pi i}{2a}}\, e^{i(a/2)(x_2-x_0)^2},$$

$$\int_R dx_2 e^{i[(a/2)(x_2-x_0)^2 + a(x_3-x_2)^2]} = \sqrt{\frac{\pi i}{(3/2)a}}\, e^{i(a/3)(x_3-x_0)^2},$$

$$\vdots$$

$$\int_R dx_{n-1} e^{i[[(a/(n-1)](x_{n-1}-x_0)^2 + a(x_n-x_{n-1})^2]} = \sqrt{\frac{\pi i}{[n/(n-1)]a}}\, e^{i(a/n)(x_n-x_0)^2}.$$

All together,

$$\int_R dx_{n-1} \cdots \int_R dx_1 \exp\left(a\left[(x_1-x_0)^2 + (x_2-x_1)^2 + \cdots + (x_n-x_{n-1})^2\right]\right)$$

$$= \sqrt{\frac{\pi i}{2a}\frac{\pi i}{(3/2)a} \cdot \ldots \cdot \frac{\pi i}{[n/(n-1)]a}}\, e^{i(a/n)(x_n-x_0)^2}$$

$$= \sqrt{\frac{i^{n-1}\pi^{n-1}}{a^{n-1}n}}\, e^{i(m/2\hbar t)(x-x_0)^2}, \qquad \text{with } a = \frac{m}{2\hbar(t/n)}.$$

Part Three. For $x_n = x$, show

$$\psi\big((0, x_0); (t, x)\big) = \lim_{n \to \infty} \left(\frac{2\pi i\hbar}{m} \frac{t}{n}\right)^{n/2} \int_{\mathbb{R}} dx_{n-1} \cdots \int_{\mathbb{R}} dx_1$$
$$\cdot \exp\left(\frac{im}{2\hbar(t/n)} \sum_{1}^{n}(x_k - x_{k-1})^2\right)$$
$$= \sqrt{\frac{m}{2\pi i\hbar t}} \, e^{(im/2\hbar)(x-x_0)^2/t}.$$

The limit is trivial.

Exercise 10.3.1. Show that

$$\psi\big((0, x_0); (t, x)\big) = \sqrt{\frac{m}{2\pi i\hbar t}} \, e^{(im/2\hbar)[(x-x_0)^2/t]}$$

is a solution of the Schrödinger Equation ($V = 0$),

$$\frac{\partial \psi}{\partial t} = \frac{i\hbar}{2m} \frac{\partial^2 \psi}{\partial x^2}, \qquad \text{with } -\infty < x < \infty, t > 0.$$

If we add the boundary condition that $\psi(0, x) = f(x)$, from the principle of superposition and from Exercise 10.3.1 we reach a formal conclusion:

$$\psi(t, x) = \int_{\mathbb{R}} \psi\big((0, x_0); (t, x)\big) f(x_0) \, dx_0, \qquad \text{for } \psi(0, x) = f(x),$$

with

$$\psi\big((0, x_0); (t, x)\big) = \left(\frac{m}{2\pi i\hbar t}\right)^{1/2} e^{(im/2\hbar t)(x-x_0)^2}.$$

That is,

$$\psi(t, x) = \left(\frac{m}{2\pi i\hbar t}\right)^{1/2} \int_{\mathbb{R}} e^{(im/2\hbar t)(x-x_0)^2} f(x_0) \, dx_0, \qquad \text{for } \psi(0, x) = f(x).$$

The reader may verify that formally

$$\frac{\partial \psi}{\partial t} = \frac{i\hbar}{2m} \frac{\partial^2 \psi}{\partial x^2}.$$

However, the integral may not make sense unless some restrictions are imposed on f.

10.4   The Fourier Transform

The Fourier transform is an important method of solving constant coefficient partial differential equations. Such methods will be helpful in solving Schrödinger Problem A. Let us briefly review such techniques.

Definition 10.4.1 (Fourier Transform). Suppose that f is a complex-valued function on R, where $\int_R |f(x)|\,dx < \infty$. In other words, f belongs to $L_C^1(R)$. Then the Fourier transform of f, $\mathcal{F}(f)$, is given by

$$\mathcal{F}(f)(y) = (2\pi)^{-1/2} \int_R e^{-ixy} f(x)\,dx.$$

Exercise 10.4.1. Let $f(x) = e^{-|x|}$, with $-\infty < x < +\infty$. Using the Lebesgue Dominated Convergence Theorem 6.3.3, show that $\mathcal{F}(f)(y) = (2\pi)^{-1/2}2(1 + y^2)^{-1}$. Note: $\mathcal{F}(f)(y)$ belongs to $L_C^1(R)$.

Exercise 10.4.2. Let

$$f(x) = \begin{cases} 1 & -1 \le x \le 1, \\ 0 & \text{otherwise.} \end{cases}$$

Show that

$$\mathcal{F}(f)(y) = (2\pi)^{-1/2}\frac{2\sin(y)}{y}.$$

Note: $\mathcal{F}(f)$ is not a member of $L_C^1(R)$, but is a member of $L_C^2(R)$.

Aside: Show that e^{iz}/z is analytic for $z \ne 0$, and conclude that the contour integral $\oint e^{iz}/z\,dz = 0$ when C is the contour; see Figure 4.

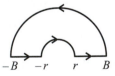

Figure 4. Contour C

Show that $\int_R \sin(y)/y\,dy = \pi$.
By the way,

$$\sum_{-\infty}^{\infty} \frac{\sin(k)}{k} = \int_{-\infty}^{\infty} \frac{\sin(x)}{x}\,dx = \int_{-\infty}^{\infty} \left(\frac{\sin(x)}{x}\right)^2 dx = \sum_{-\infty}^{\infty} \left(\frac{\sin(k)}{k}\right)^2.$$

Exercise 10.4.3. Let

$$f(x) = \begin{cases} e^{-x} & x \geq 0, \\ 0 & x < 0. \end{cases}$$

Show that

$$\mathcal{F}(f)(y) = (2\pi)^{-1/2} \left(\frac{1 - iy}{1 + y^2} \right).$$

This Fourier transform is complex-valued.

Exercise 10.4.4. Let $f(x) = e^{-x^2/2}$, where $-\infty < x < +\infty$. Argue that $\mathcal{F}(f)(y)$ is a differentiable function of y and show $\left(\mathcal{F}(f)(y) \right)' + y\mathcal{F}(f)(y) = 0$, with $\mathcal{F}(f)(0) = 1$. Solve this differential equation, and conclude that $\mathcal{F}\left(e^{-x^2/2} \right)(y) = e^{-y^2/2}$.

Note that in the preceding exercise $y^2 \mathcal{F}\left(e^{-x^2/2} \right)$ belongs to $L_C^2(\mathbb{R})$.

Exercise 10.4.5. Show $\mathcal{F}\left(e^{(-s/2)x^2} \right)(y) = s^{-1/2}e^{-y^2/2s}$, for $s > 0$. Hint: $\mathcal{F}'$.

Exercise 10.4.6. We want to replace real s in Exercise 10.4.5 with complex z having positive real part. That is, we want

$$(2\pi)^{-1/2} \int_{\mathbb{R}} e^{-ixy} e^{(-z/2)x^2} \, dx = z^{-1/2} e^{-y^2/2z},$$

for $z = a + ib$, $a > 0$. Show that $(2\pi)^{-1/2} \int_{\mathbb{R}} e^{-ixy} e^{(-z/2)x^2} \, dx$ and $z^{-1/2} e^{-y^2/2z}$ are analytic (Cauchy–Riemann equations). Conclude

$$\mathcal{F}\left(e^{(-z/2)x^2} \right)(y) = z^{-1/2} e^{-y^2/2z}$$

for Re $z > 0$.

10.5   The Convolution Product

We will need a special product of functions, the *convolution product*, to solve Schrödinger Problem A.

Definition 10.5.1 (Convolution Product). Suppose f and g are members of $L_C^2(\mathbb{R})$. The convolution product of f with g, $f * g$ is

$$(f * g)(x) \equiv (2\pi)^{-1/2} \int_{\mathbb{R}} f(y)g(x - y) \, dy.$$

Does this product make sense? Yes: by the Hölder–Riesz Inequality (Theorem 6.5.2),

$$\int_R |f(y)g(x - y)| \, dy \leq \left(\int_R |f|^2 \, dx \right)^{1/2} \cdot \left(\int_R |g|^2 \, dx \right)^{1/2}.$$

Suppose f and g are members of $L_C^1(R)$. Does the convolution product $f * g$ still make sense? We see that

$$\int_R \left(\int_R |f(y)g(x - y)| \, dx \right) dy = \int_R |f(y)| \left(\int_R |g(x - y)| \, dx \right) dy$$
$$= \int_R |f(y)| \left(\int_R |g(u)| \, du \right) dy$$
$$= \int_R |f(y)| \, dy \int_R |g(u)| \, du < \infty.$$

Thus $f(y)g(x - y)$ belongs to $L_C^1(R)$.

By Fubini's Theorem, $\int_{-\infty}^{\infty} f(y)g(x - y) \, dy$ exists for almost all x and is integrable. So $f * g$ makes sense for f and g members of $L_C^1(R)$.

10.6   The Schwartz Space

For Schrödinger Problem A we will also find useful the inverse Fourier transforms, and the Schrödinger requirement $\int_R |\psi(t, x)|^2 \, dx = 1$, for $t \geq 0$, suggests pleasant behavior for large values of x. This brings us to the Schwartz space.

Definition 10.6.1 (Schwartz Space). The Schwartz space $\mathcal{S}$ consists of those complex-valued functions on R, $f : R \rightarrow \mathcal{C}$, such that

1. f has derivatives of all orders.

2. f and all its derivatives decrease to zero as $|x| \rightarrow \infty$:

$$\lim_{|x| \to \infty} |x|^m D^n f(x) = 0$$

for all nonnegative integers m and n.

It can be shown that the Schwartz space is a dense subspace of $L_C^p(R)$ for $p \geq 1$.

The Fourier transform on the Schwartz space has many beautiful properties.

Exercise 10.6.1.

a. Show that $\mathcal{F}(f)(y) = (2\pi)^{-1/2} \int_R e^{-ixy} f(x)\,dx$ for $f \in \mathcal{S}$ has derivatives of all orders. In particular, show that $\big(\mathcal{F}(f)(y)\big)^{(n)} = (-i)^n \mathcal{F}(x^n f)(y)$. Show also that $\mathcal{F}\big(f^{(n)}\big)(y) = (-iy)^n \mathcal{F}(f)(y)$.

b. Show $\mathcal{F}$ maps $\mathcal{S}$ into $\mathcal{S}$, or $\mathcal{F}(\mathcal{S}) \subseteq \mathcal{S}$.

10.6.1   Plancherel's Theorem

We have a useful theorem due to Michel Plancherel, whose proof can be found in Weidmann (1980, p. 292). The theorem has five parts.

1. The Fourier transform is a one-to-one linear map of the Schwartz space to itself. In fact, the inverse Fourier transform of $\mathcal{F}$, $\mathcal{F}^{-1}$, is given by

$$\mathcal{F}^{-1}(f)(x) = (2\pi)^{-1/2} \int_R e^{ixy} f(y)\,dy$$

   for f a member of the Schwartz space $\mathcal{S}$. Note that $\mathcal{F}^{-1}(f)(y) = \mathcal{F}(f)(-y)$.

2. For all members of the Schwartz space $\mathcal{S}$, a dense subspace of $L^2_{\mathcal{C}}(R)$,

$$\|\mathcal{F}(f)\|^2 = \|f\|^2 = \|\mathcal{F}^{-1}(f)\|^2.$$

3. The Fourier transform $\mathcal{F}$ and its inverse $\mathcal{F}^{-1}$ can be extended from the Schwartz space $\mathcal{S}$ to all of $L^2_{\mathcal{C}}(R)$. If f is a member of $L^2_{\mathcal{C}}(R)$, then for $r > 0$,

$$\mathcal{F}_r(f)(y) = (2\pi)^{-1/2} \int_{-r}^{r} e^{-ixy} f(x)\,dx$$

   is a member of $L^2_{\mathcal{C}}(R)$. The mean-square limit — which is called the *Fourier transform* of f and is also written $\mathcal{F}(f)(y)$ — exists and defines a function in $L^2_{\mathcal{C}}(R)$:

$$\|\mathcal{F}(f)(y) - \mathcal{F}_r(f)(y)\|^2 \to 0 \text{ as } r \to \infty.$$

   That is, for $f \in L^2_{\mathcal{C}}(R)$,

$$\mathcal{F}(f)(y) \equiv \lim_{r \to \infty} (2\pi)^{-1/2} \int_{-r}^{r} e^{-ixy} f(x)\,dx$$

$$\equiv (2\pi)^{-1/2} \int_R e^{-ixy} f(x)\,dx.$$

Similarly,

$$\mathcal{F}^{-1}(f)(x) = \lim_{r \to \infty} (2\pi)^{-1/2} \int_{-r}^{r} e^{ixy} f(y) \, dy$$
$$\equiv (2\pi)^{-1/2} \int_{R} e^{ixy} f(y) \, dy.$$

4. For all members f of $L_{\mathcal{C}}^2(R)$, the Fourier and inverse Fourier trans-
forms are bounded linear operators with $\|\mathcal{F}(f)\|^2 = \|f\|^2 = \|\mathcal{F}^{-1}(f)\|^2$.

5. For f and g members of $L_{\mathcal{C}}^2(R)$, the convolution product $f * g$ has
the property

$$(2\pi)^{-1/2} f * g = \mathcal{F}^{-1}\left(\mathcal{F}(f) \cdot \mathcal{F}(g)\right) = \mathcal{F}\left(\mathcal{F}^{-1}(f) \cdot \mathcal{F}^{-1}(g)\right).$$

Moreover, the inner product $\langle f, g \rangle \equiv \int_R f \bar{g} \, dx$ satisfies

$$\langle f, g \rangle = \langle \mathcal{F}(f), \mathcal{F}(g) \rangle = \langle \mathcal{F}^{-1}(f), \mathcal{F}^{-1}(g) \rangle.$$

Exercise 10.6.2. Using Plancherel's Theorem, show that

$$e^{-(z/2)y^2} = \mathcal{F}\left(z^{-1/2} e^{-x^2/2z}\right)(y)$$

for Re $z > 0$. Hint: By Exercise 10.4.6,

$$\mathcal{F}\left(e^{(-z/2)x^2}\right)(y) = z^{-1/2} e^{-y^2/2z},$$

for Re $z > 0$.

As an application of Plancherel's Theorem, we showed (Exercise 10.4.2)
that the Fourier transform of

$$f(x) = \begin{cases} 1 & -1 \le x \le 1, \\ 0 & \text{otherwise,} \end{cases}$$

is given by

$$\mathcal{F}(f)(y) = (2\pi)^{-1/2} \frac{2 \sin(y)}{y}.$$

So f is a member of $L_{\mathcal{C}}^2(R) \cap L_{\mathcal{C}}^1(R)$, and $\mathcal{F}(f)$ is a member of $L_{\mathcal{C}}^2(R) \setminus L_{\mathcal{C}}^1(R)$.

We have

$$\mathcal{F}^{-1}(\mathcal{F}(f)(y)) = \lim_{r\to\infty} (2\pi)^{-1/2} \int_{-r}^{r} e^{ixy}\mathcal{F}(f)(y)\,dy$$

$$= \lim_{r\to\infty} (2\pi)^{-1} \int_{-r}^{r} e^{ixy}\frac{2\sin(y)}{y}\,dy$$

$$= (2\pi)^{-1} \lim_{r\to\infty} \int_{-r}^{r} e^{ixy}\left(\frac{e^{iy}-e^{-iy}}{iy}\right)\,dy$$

$$= (2\pi)^{-1} \lim_{r\to\infty} \int_{-r}^{r} \frac{\sin\left((x+1)y\right)-\sin\left((x-1)y\right)}{y}\,dy$$

$$= \begin{cases} 1 & -1 < x < 1, \\ \frac{1}{2} & x = \pm 1, \\ 0 & \text{otherwise,} \end{cases}$$

$$= f \text{ almost everywhere.}$$

Exercise 10.6.3. We showed in Exercise 10.4.1 that the Fourier transform of $e^{-|x|}$ is given by $(2\pi)^{-1/2}2(1+y^2)^{-1}$. Show that

$$e^{-|x|} = (2\pi)^{-1/2} \lim_{r\to\infty} \int_{-r}^{r} (2\pi)^{-1/2}2(1+y^2)^{-1}e^{ixy}\,dy$$

$$= \frac{1}{\pi} \int_{-\infty}^{\infty} \frac{\cos(xy)}{1+y^2}\,dy.$$

In particular, for $x = 0$,

$$1 = \frac{1}{\pi} \int_{-\infty}^{\infty} \frac{1}{1+y^2}\,dy$$

$$= \frac{1}{\pi} \lim_{r\to\infty} \left[\tan^{-1}(r) - \tan^{-1}(-r)\right]$$

Exercise 10.6.4. Suppose that $f(x) = x/(1+x^2)$, where $-\infty < x < +\infty$.

a. Show that f is not a member of $L_C^1(\mathbf{R})$.

b. Show that f is a member of $L_C^2(\mathbf{R})$.

c. Calculate $\mathcal{F}(f)(y)$. Hint: See Figure 5; $\int_C z/(1+z^2)e^{-izy}\,dz$.

d. Is $y^2\mathcal{F}(f)(y)$ a member of $L_C^2(\mathbf{R})$?

Exercise 10.6.5. Suppose $f(x) = 1/(1+x^2)$, for $-\infty < x < +\infty$. Calculate $\mathcal{F}(f)$.

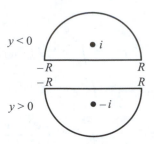

Figure 5. Contour C

10.7 Solving Schrödinger Problem A

Using Fourier transforms, let's solve (heuristically) the Schrödinger Equation in Problem A in the form $(V = 0)$,

$$\frac{\partial \psi}{\partial t} = \frac{i\hbar}{2m} \frac{\partial^2 \psi}{\partial x^2}, \qquad \text{for } -\infty < x < \infty, t > 0,$$

$$\psi(0, x) = f(x), \qquad \text{with } \int_{\mathbb{R}} |\psi(t, x)|^2 \, dx = 1, t \geq 0.$$

Assume $\psi(t, x)$ has a Fourier transform:

$$\mathcal{F}\big(\psi(t, x)\big)(y) = (2\pi)^{-1/2} \int_{\mathbb{R}} e^{-ixy} \psi(t, x) \, dx.$$

If ψ and f are nice enough, then

$$\mathcal{F}\left(\frac{\partial \psi}{\partial t}\right) = \frac{\partial}{\partial t} \mathcal{F}, \ \mathcal{F}\left(\frac{\partial^2 \psi}{\partial x^2}\right) = -y^2 \mathcal{F}(\psi), \ \mathcal{F}\big(\psi(0, x)\big)(y) = \mathcal{F}(f)(y),$$

and thus

$$\frac{\partial \mathcal{F}}{\partial t} = \frac{-i\hbar}{2m} y^2 \mathcal{F}.$$

We conclude that $\mathcal{F}(\psi) = e^{(-i\hbar/2m)y^2 t} \mathcal{F}(f)$. Then

$$\begin{aligned}
\psi(t, x) &= \mathcal{F}^{-1}\big(e^{(-i\hbar/2m)y^2 t} \mathcal{F}(f)\big)(x) \\
&= \mathcal{F}^{-1}\big(e^{(-i\hbar/2m)y^2 t}\big)(x) * \mathcal{F}^{-1}\big(\mathcal{F}(f)\big)(x) \\
&= \left(\frac{i\hbar t}{2m}\right)^{-1/2} e^{-x^2/[2i\hbar(t/m)]} * f(x) \\
&= \left(\frac{m}{2\pi i\hbar t}\right)^{1/2} \int_{\mathbb{R}} e^{(im/2\hbar t)(x-x_0)^2} f(x_0) \, dx_0.
\end{aligned}$$

Note how the third line recalls Exercises 10.4.6 and 10.6.2.

We have several problems. For instance:

- What does $\mathcal{F}^{-1}\left(e^{(-i\hbar/2m)y^2 t}\right)$ mean when Re $z = 0$? Note that $e^{-i\hbar/2my^2 t}$ is not in $L_C^1(\mathrm{R})$.

- Are there any requirements on f besides $\int_\mathrm{R} |f(x)|^2 \, dx = 1$?

Ignoring these issues for the present, let's compare this expression (via Fourier transforms) with Feynman's expression (via Riemann sums) as a solution of the Schrödinger Equation,

$$\frac{\partial \psi}{\partial t} = \frac{i\hbar}{2m} \frac{\partial^2 \psi}{\partial x^2}, \qquad \text{for } -\infty < x < \infty, t > 0,$$

$$\psi(0, x) = f(x), \quad \text{with} \int_\mathrm{R} |\psi(t, x)|^2 \, dx = 1, t \geq 0.$$

That is, we compare

$$\psi(t, x) = \left(\frac{m}{2\pi i\hbar t}\right)^{1/2} \int_\mathrm{R} \exp\left\{\frac{im}{2\hbar t}(x - x_0)^2\right\} f(x_0) \, dx_0$$

with Feynman's expression

$$\psi(t, x) = \lim_n \left(\frac{m}{2\pi i\hbar(t/n)}\right)^{n/2} \int_\mathrm{R} dx_{n-1} \cdots \int_\mathrm{R} dx_0$$

$$\cdot \exp\left\{\frac{im}{2\hbar(t/n)} \sum_1^n (x_k - x_{k-1})^2\right\} f(x_0),$$

where $x_n = x$, $\psi(0, x) = f(x)$. The common features of the integrands are encouraging.

10.7.1 Finding the Right Space of Functions

The requirement that $\int_\mathrm{R} |\psi(t, x)|^2 \, dx = 1$ for $t \geq 0$ suggests that we are looking for a solution of the Schrödinger Equation in the space of complex-valued square integrable functions on R — the Hilbert space $L_C^2(\mathrm{R})$.

On the other hand, we have differentiability requirements:

- What does it mean to say we have a solution of Schrödinger's Equation in $L_C^2(\mathrm{R})$?

- Is there a dense subspace of $L_C^2(\mathrm{R})$ where differentiability makes sense?

- Do the elements of this subspace (and their derivatives) satisfy growth restrictions as $x \to \pm\infty$?

We want a dense subspace of $L^2_{\mathbb{C}}(\mathbb{R})$ consisting of smooth functions that decay sufficiently fast. The Schwartz space is our space.

10.8   An Abstract Cauchy Problem

We are going to reformulate the Schrödinger problem as an abstract Cauchy problem, an approach developed by Einar Hille and Ralph S. Phillips in the 1940s and 1950s. Our discussion follows Fattorini (1983), Goldstein (1985), and Johnson and Lapidus (2000).

Think of $u(t)$ as the state of a physical system u at time t. Think of the time rate of change of $u(t)$, $u'(t)$, as a function A of the state of the system u. We are given the initial datum $u(0) = f$. We have $u'(t) = A[u]$, with $u(0) = f$. Thus the Schrödinger Equation for $V = 0$,

$$\frac{\partial \psi}{\partial t} = \frac{i\hbar}{2m} \frac{\partial^2 \psi}{\partial x^2}, \qquad \text{for } -\infty < x < \infty, t > 0,$$

$$\psi(0, x) = f(x), \qquad \text{with } -\infty < x < \infty, \int_{\mathbb{R}} |\psi(t, x)|^2 \, dx = 1, \ t \geq 0,$$

can now be reformulated as a differential equation in the Hilbert space $L^2_{\mathbb{C}}(\mathbb{R})$. We have

$$u'(t) = A[u(t)], \qquad \text{for } t \geq 0;$$
$$u(0) = f.$$

The differential operator A, with domain an appropriate dense subspace of $L^2_{\mathbb{C}}(\mathbb{R})$, is given by the differential expression

$$\frac{i\hbar}{2m} \frac{d^2}{dx^2},$$

while $u(0)(x) = f(x)$ and $u(t)(x) = \psi(t, x)$.

10.8.1   Defining the Abstract Cauchy Problem

In the abstract Cauchy problem we have

$$u'(t) = A[u(t)], \qquad \text{for } t \geq 0,$$
$$u(0) = f,$$

with the following two conditions:

1. The differential operator A is a linear operator from a Banach space X to itself. The domain of A is a dense subspace of X.

2. The solution of this differential equation $u'(t) = A[u(t)]$ for $t \geq 0$ is a function $u(t)$ continuously differentiable for $t \geq 0$.

By Condition 2 we mean that

$$u'(t) = \lim_{h \to 0} \frac{u(t+h) - u(t)}{h}$$

exists and is continuous in the norm of X. The function $u(t)$ belongs to the domain of A, with

$$\lim_{h \to 0} \left\| \frac{u(t+h) - u(t)}{h} - A[u(t)] \right\|_X = 0.$$

The Cauchy problem is well posed if the following two conditions are satisfied:

1. We have a dense subspace of X such that for any member u_0 of this dense subspace we have a solution $u'(t) = A[u(t)]$ with $u(0) = u_0$.

2. We have a nondecreasing, nonnegative function $B(t)$ defined for $t \geq 0$ such that, for any solution v of $u'(t) = A[u(t)]$,

$$\|v(t)\|_X \leq B(t) \|v(0)\|_X ,$$

for $t \geq 0$.

10.8.2 Operators on a Complex Hilbert Space

It will be helpful to consider some general comments regarding operators on a complex Hilbert space $\mathcal{H}$. The domain of a linear operator T — that is, $D(T)$ — is a subspace. We have $T(\alpha_1 x_1 + \alpha_2 x_2) = \alpha_1 T x_1 + \alpha_2 T x_2$ for all scalars α_1, α_2 and all elements x_1, x_2 in $D(T)$. The following terminology applies.

1. Bounded. A linear operator T in $\mathcal{H}$ is *bounded* if there exists a constant $C \geq 0$ so that $\|Tf\| \leq C \|f\|$ for all f in the domain of T.

2. Continuous. Saying a linear operator is *continuous* at f in $D(T)$ means that for every sequence $\{f_n\}$ in $D(T)$ for which $\|f_n - f\| \to 0$ it follows that $\|Tf_n - Tf\| \to 0$.

3. Bounded iff Continuous. A linear operator T is bounded iff it is continuous. If T is bounded, $\|Tf\| \leq C\|f\|$, then $\|Tf_n - Tf\| = \|T(f_n - f)\| \leq C\|f_n - f\|$.

But if T is not bounded, we have $\{f_n\}$ in $D(T)$, so that $\|Tf_n\| \geq n\|f_n\|$. For $g_n = (1/n\|f_n\|)f_n$, $\|g_n\| \to 0$, but $\|Tg_n\| \geq 1$.

4. Norm. The norm of T, $\|T\|$, is defined by

$$\|T\| \equiv \sup_{\|f\|=1} \|Tf\| = \sup \frac{\|Tf\|}{\|f\|} \qquad \text{for } f \text{ a member of } D(T).$$

5. Unitary. A bounded linear operator U of $\mathcal{H}$ into $\mathcal{C}$ is *unitary* iff U is isometric, $\langle Uf, Uf \rangle = \langle f, f \rangle$, and onto. For example, the Fourier transform $\mathcal{F}$ is a unitary operator, by Plancherel's Theorem.

In all that follows, $\mathcal{H}$ will be the Hilbert space of complex-valued Lebesgue measurable functions defined on R, such that $|f|^2$ is Lebesque integrable: $\int_R |f|^2 \, dx < \infty$. In this space the inner product of functions f and g is defined by

$$\langle f, g \rangle \equiv \int_D \int_R f\overline{g} \, dx,$$

$$\|f\|^2 = \langle f, f \rangle = \int_R |f|^2 \, dx.$$

We follow tradition by denoting this space as $L^2_{\mathcal{C}}(R)$. Convergence of f_n to f means that

$$\|f_n - f\| = \left(\int_R |f_n - f|^2 \, dx \right)^{1/2} \to 0 \text{ as } n \to \infty.$$

The space $L^2_{\mathcal{C}}(R)$ is a complete metric space (see Chapter 6).

10.9   Solving in the Schwartz Space

Example 10.9.1. Given Problem A, where $V = 0$ in the Schrödinger Equation,

$$\frac{\partial \psi}{\partial t} = \frac{i\hbar}{2m} \frac{\partial^2 \psi}{\partial x^2}, \qquad \text{for } -\infty < x < \infty, t > 0,$$

$$\psi(0, x) = f(x), \qquad \text{with } \int_R |\psi(t, x)|^2 \, dx = 1, \ t \geq 0,$$

we will solve the corresponding abstract Cauchy problem $u'(t) = A[u(t)]$, with $u(0) = f$. Suppose f is a member of the Schwartz space $\mathcal{S}$, a dense subspace of $L^2_C(R)$, $\int_R |f(x)|^2\, dx = 1$, and the domain of the differential operator

$$A = \frac{i\hbar}{2m}\frac{d^2}{dx^2}$$

is the Schwartz space S.

This is a lengthy exploration, and we will break it into three parts.

Part One. Assuming u is a member of S, the Fourier transform of u is given by

$$\mathcal{F}\big(u(t)(x)\big)(y) = (2\pi)^{-1/2}\int_R e^{-ixy}u(t)(x)\, dx.$$

Proceeding as in Section 10.7, we have

$$\big(\mathcal{F}(u)\big)' = \mathcal{F}(u') = \mathcal{F}(A[u]) = \frac{-i\hbar}{2m}y^2\mathcal{F}(u),$$

with $\mathcal{F}\big(u(0)(x)\big) = \mathcal{F}(f(x))$. Solving this differential equation yields $\mathcal{F}(u) = e^{(-i\hbar/2m)ty^2}\mathcal{F}(f)$. Because $e^{(-i\hbar/2m)ty^2}\mathcal{F}(f)$ is a member of S, we may conclude that

$$u(t)(x) = \mathcal{F}^{-1}\left(e^{(-i\hbar/2m)y^2 t}\mathcal{F}(f)(y)\right)(x), \qquad \text{with } u(0)(x) = f(x).$$

Also, $u(t)(x)$ is a member of S, the domain of A, by Plancherel's Theorem (Section 10.6.1).

The reader should verify that

$$\psi(t,x) \equiv u(t)(x) = \mathcal{F}^{-1}\left(e^{(-i\hbar/2m)y^2 t}\mathcal{F}(f)(y)\right)(x)$$

solves the Schrödinger Equation with $V = 0$. Hint: By Plancherel's Theorem, $\psi(t,x)$ is a member of S. That is,

$$\psi(t,x) = (2\pi)^{-1/2}\int_R e^{ixy}\left[e^{(-i\hbar/2m)y^2 t}\mathcal{F}(f)(y)\right]dy$$

is a member of S. Show that

$$\frac{\partial\psi}{\partial t} = (2\pi)^{-1/2}\int_R \frac{-i\hbar y^2}{2m}e^{ixy}\left[e^{(-i\hbar/2m)y^2 t}\mathcal{F}(f)(y)\right]dy \qquad \text{and}$$

$$\frac{\partial^2\psi}{\partial x^2} = (2\pi)^{-1/2}\int_R -y^2 e^{ixy}\left[e^{(-i\hbar/2m)y^2 t}\mathcal{F}(f)(y)\right]dy.$$

Furthermore, $\psi(0, x) = \mathcal{F}^{-1}\big(\mathcal{F}(f)(y)\big)(x) = f(x)$ and

$$
\begin{aligned}
\int_{\mathbb{R}} |\psi(t, x)|^2 \, dx &= \langle \psi(t, x), \psi(t, x) \rangle \\
&= \Big\langle \mathcal{F}^{-1}\big(e^{(-i\hbar/2m)y^2 t} \mathcal{F}(f)\big), \big(\mathcal{F}^{-1} e^{(-i\hbar/2m)y^2 t} \mathcal{F}(f)\big) \Big\rangle \\
&= \Big\langle e^{(-i\hbar/2m)y^2 t} \mathcal{F}(f), e^{(-i\hbar/2m)y^2 t} \mathcal{F}(f) \Big\rangle \\
&= \langle \mathcal{F}(f), \mathcal{F}(f) \rangle = \langle f, f \rangle \\
&= \int_{\mathbb{R}} |f(x)|^2 \, dx = 1 \qquad \text{for } t \geq 0.
\end{aligned}
$$

Part Two. Next we will verify that, given

$$
u(t)(x) = \mathcal{F}^{-1}\big(e^{(-i\hbar/2m)y^2 t} \mathcal{F}(f)(y)\big)(x),
$$

u solves the abstract Cauchy problem in the sense of our definition (Section 10.8.1). Assume f is a member of $\mathcal{S}$. By Plancherel's Theorem, u is a member of $\mathcal{S}$ and we can calculate

$$
\begin{aligned}
&\frac{u(t + h) - u(t)}{h} \\
&= \mathcal{F}^{-1}\left(\frac{e^{(-i\hbar/2m)y^2(t+h)} - e^{(-i\hbar/2m)y^2 t}}{h} \mathcal{F}(f)(y) \right) \\
&= \mathcal{F}^{-1}\left(e^{(-i\hbar/2m)y^2(t+h/2)} \cdot \left(\frac{-2i \sin\left(\frac{\hbar}{2m} y^2 \frac{h}{2}\right)}{\frac{\hbar}{2m} y^2 \frac{h}{2}} \right) \cdot \frac{\hbar}{2m} \frac{y^2}{2} \cdot \mathcal{F}(f)(y) \right) \\
&\xrightarrow[h \to 0]{} \mathcal{F}^{-1}\left(e^{(-i\hbar/2m)y^2 t} \cdot \frac{-i\hbar}{2m} y^2 \mathcal{F}(f)(y) \right) \\
&= \mathcal{F}^{-1}\left(\frac{-i\hbar}{2m} y^2 \mathcal{F}(u(t)) \right) = \mathcal{F}^{-1}\big(\mathcal{F}(A[u(t)])\big) = A[u(t)].
\end{aligned}
$$

Also, for any solution v,

$$
\begin{aligned}
\langle A[v], v \rangle &= \langle \mathcal{F}(A[v]), \mathcal{F}(v) \rangle = \left\langle \frac{-i\hbar}{2m} y^2 \mathcal{F}(v), \mathcal{F}(v) \right\rangle \\
&= \frac{-i\hbar}{2m} \int_{\mathbb{R}} y^2 |\mathcal{F}(v)|^2 \, dy.
\end{aligned}
$$

Thus Re $\langle A[v], v \rangle = 0$.

Because

$$\frac{d}{dt}\|v\|^2 = 2\mathrm{Re}\left\langle\frac{dv}{dt}, v\right\rangle = 2\mathrm{Re}\,\langle A[v], v\rangle = 0,$$

we conclude that $\|v\|^2 = \|v(0)\|^2 = \int_{\mathrm{R}} |f|^2\,dx = 1$, for $t \geq 0$. So for f a member of $\mathcal{S}$, $\mathcal{F}^{-1}\left(e^{(-i\hbar/2m)y^2 t}\mathcal{F}(f)\right)(x)$ solves the abstract Cauchy problem.

Part Three. How does this solution compare with our heuristic solution in Section 10.7? Recall that we arrived at

$$\left(\frac{m}{2\pi i\hbar t}\right)^{1/2} \int_{\mathrm{R}} \exp\left(\frac{im}{2\hbar t}(x-x_0)^2\right) f(x_0)\,dx_0.$$

We need an integral representation for $\mathcal{F}^{-1}\left(e^{(-i\hbar/2m)y^2 t}\mathcal{F}(f)(y)\right)(x)$.

From Plancherel's Theorem (Section 10.6.1) and Exercise 10.6.2 we know that if f is a member of $L^2_{\mathbb{C}}(\mathrm{R})$, then for $\mathrm{Re}\,z > 0$,

$$\mathcal{F}^{-1}\left(e^{(-z/2)y^2}\mathcal{F}(f)(y)\right)(x) = \mathcal{F}^{-1}\left(\mathcal{F}(z^{-1/2}e^{-x^2/2z})(y) \cdot \mathcal{F}(f)(y)\right)(x)$$

$$= (2\pi)^{-1/2}z^{-1/2}e^{-x^2/2z} * f(x)$$

$$= (2\pi)^{-1/2}\int_{\mathrm{R}} e^{-(x-x_0)^2/2z} f(x_0)\,dx_0,$$

for $\mathrm{Re}\,z > 0$.

Select a sequence $z_n \to (i\hbar/m)t$ for $t > 0$, with $\mathrm{Re}\,(z_n) > 0$. Because $\mathcal{F}^{-1}$ is a bounded linear operator on the complete normed linear space $L^2_{\mathbb{C}}(\mathrm{R})$ (by Plancherel's Theorem), it follows that

$$\mathcal{F}^{-1}\left(e^{(-z_n/2)y^2}\mathcal{F}(f)(y)\right) \to \mathcal{F}^{-1}\left(e^{(-i\hbar/2m)y^2 t}\mathcal{F}(f)(y)\right)(x)$$

in $L^2_{\mathbb{C}}(\mathrm{R})$, and we have subsequence $\{z_{n_k}\}$ such that pointwise

$$\lim_k \mathcal{F}^{-1}\left(e^{(-z_{n_k}/2)y^2}\mathcal{F}(f)(y)\right)(x) = \mathcal{F}^{-1}\left(e^{(-i\hbar/2m)y^2 t}\mathcal{F}(f)(y)\right)(x)$$

almost everywhere in x (Theorem 6.5.4).

Assuming f is a member of $L^1_{\mathbb{C}}(\mathrm{R}) \cap L^2_{\mathbb{C}}(\mathrm{R})$, the Lebesgue Dominated

Convergence Theorem 6.3.3 gives us

$$\mathcal{F}^{-1}\left(e^{(-i\hbar/2m)y^2 t}\mathcal{F}(f)(y)\right)(x)$$

$$= \lim_k \mathcal{F}^{-1}\left(e^{(-z_{n_k}/2)y^2} \cdot \mathcal{F}(f)(y)\right)(x)$$

$$= \lim_k \mathcal{F}^{-1}\left(\mathcal{F}\left(z_{n_k}^{-1/2}e^{-x^2/2z_{n_k}}\right)(y) \cdot \mathcal{F}(f)(y)\right)(x)$$

$$= \lim_k z_{n_k}^{-1/2}e^{-x^2/2z_{n_k}} * f(x)$$

$$= \lim_k \left(2\pi z_{n_k}\right)^{-1/2}\int_{R} e^{-(x-x_0)^2/2z_{n_k}} f(x_0)\, dx_0$$

$$= \left(\frac{m}{2\pi i\hbar t}\right)^{1/2}\int_{R} e^{(im/2\hbar t)(x-x_0)^2} f(x_0)\, dx_0$$

almost everywhere in x. Note that for $\mathrm{Re}\, z \geq 0$, $z \neq 0$, $\left|e^{-(x-x_0)^2/2z}\right| \leq 1$, with f a member of

$$\mathcal{F}^{-1}\left(e^{(-i\hbar/2m)y^2 t}\mathcal{F}(f)(y)\right)(x)$$

$$= \left(\frac{m}{2\pi i\hbar t}\right)^{1/2}\int_{R} e^{(im/2\hbar t)(x-x_0)^2} f(x_0)\, dx_0,$$

almost everywhere in x, with the assumption that f is a member of $L_C^1(R)\cap L_C^2(R)$. (The integral is an ordinary Lebesgue integral.)

Since $\mathcal{S} \subset L_C^1(R) \cap L_C^2(R)$, we may conclude that the solution of the Schrödinger Equation (Section 10.1.1) or the solution of the abstract Cauchy problem (Section 10.8.1) has the two representations,

$$u(t)(x) = \mathcal{F}^{-1}\left(e^{(-i\hbar/2m)y^2 t}\mathcal{F}(f)(y)\right)(x)$$

$$= \left(\frac{m}{2\pi i\hbar t}\right)^{1/2}\int_{R} \exp\left(\frac{im}{2\hbar t}(x - x_0)^2\right) f(x_0)\, dx_0$$

almost everywhere in x, for f in $\mathcal{S}$. (The integral is an ordinary Lebesgue integral.)

10.9.1   Extending the Solution of Problem A

Let's try an extension of what we mean by a "solution" of Schrödinger Problem A, based on Johnson and Lapidus (2000, pp. 166–168).

Example 10.9.2. We will enlarge the domain of A to a subspace of $L_C^2(R)$ containing $\mathcal{S}$.

For f a member of S we showed that

$$A[f] = \frac{i\hbar}{2m} \frac{d^2 f}{dx^2} = \mathcal{F}^{-1}\left(\frac{-i\hbar}{2m} y^2 \mathcal{F}(f)(y)\right)(x).$$

However, the operator $\mathcal{F}^{-1}\left((-i\hbar/2m)y^2\mathcal{F}(f)(y)\right)$ is defined when $y^2\mathcal{F}(f)$ is a member of $L^2_C(R)$ or when $(1 + y^2)\mathcal{F}(f)$ is a member of $L^2_C(R)$. So we extend the operator A:

$$A[f] \equiv \mathcal{F}^{-1}\left((-i\hbar/2m)y^2\mathcal{F}(f)\right)$$

with domain $D(A) = \{g \in L^2_C(R) \mid (1 + y^2)\mathcal{F}(g) \in L^2_C(R)\}$. The ordinary second derivative of f, $\dfrac{d^2 f}{dx^2}$, is being replaced by the operator

$$\mathcal{F}^{-1}\left(-y^2\mathcal{F}(f)\right)$$

with the requirement that $(1 + y^2)\mathcal{F}(f)$ is a member of $L^2_C(R)$. The domain $D(A)$ is called the *Sobelev space of order* 2.

So how does this work? We have

$$\mathcal{F}\left(\frac{f(x+h) - f(x)}{h}\right)(y) = \frac{1}{h}(e^{iyh} - 1)\mathcal{F}(f)(y),$$

where $\left|1/h(e^{ihy} - 1)\right| \leq |y| < \sqrt{1 + y^2} \leq 1 + y^2$ and $1/h(e^{iyh} - 1) \to iy$ as $h \to 0$. So

$$\frac{f(x+h) - f(x)}{h} \to \mathcal{F}^{-1}\left(iy\mathcal{F}(f)(y)\right)(x) \qquad \text{in } L^2_C(R).$$

We have a subsequence $\{h_n\}$ converging to 0 so that $[f(x + h_n) - f(x)]/h_n$ converges pointwise almost everywhere in x as $h_n \to 0$. If f is a member of S, then $f'(x) = \mathcal{F}^{-1}\left(iy\mathcal{F}(f)(y)\right)(x)$. Similarly, $f''(x) = \mathcal{F}^{-1}\left(iyiy\mathcal{F}(f)(y)\right)(x)$.

Sometimes we say that $\mathcal{F}^{-1}\left(iy\mathcal{F}(f)(y)\right)$ and $\mathcal{F}^{-1}\left(-y^2\mathcal{F}(f)(y)\right)$ — $f^{(1)}$ and $f^{(2)}$, respectively — are distributed derivatives, or weak derivatives, or L^2 derivatives, and so on. They agree almost everywhere with the ordinary derivatives f' and f'' when f has such derivatives and they make sense as operators on a dense subspace of $L^2_C(R)$ that contains S, the so-called Sobelev space:

$$\{g \in L^2_C(R) \mid (1 + y^2)\,\mathcal{F}(g) \in L^2_C(R\} = D(A).$$

We note that for f in the domain of A and g an infinitely differentiable function of compact support,

$$\langle f^{(1)}, g \rangle = \langle \mathcal{F}^{-1}(iy\mathcal{F}(f)(y)), g \rangle$$
$$= \langle iy\mathcal{F}(f)(y), \mathcal{F}(g) \rangle = -\langle \mathcal{F}(f)(y), iy\mathcal{F}(g) \rangle$$
$$= -\langle f, \mathcal{F}^{-1}(iy\mathcal{F}(g)(y)) \rangle = -\langle f, g' \rangle$$

and

$$\langle f^{(2)}, g \rangle = \langle \mathcal{F}^{-1}(-y^2\mathcal{F}(f)(y)), g \rangle$$
$$= \langle f, \mathcal{F}^{-1}(-y^2\mathcal{F}(g)(y)) \rangle = \langle f, g'' \rangle.$$

That is, the "integration by parts" formulae,

$$\int_R f^{(1)}(x)g(x)\,dx = -\int_R f(x)g'(x)\,dx \qquad \text{and}$$

$$\int_R f^{(2)}(x)g(x)\,dx = \int_R f(x)g''(x)\,dx,$$

hold for all infinitely differentiable functions g of compact support. Thus $u(t)(x) = \mathcal{F}^{-1}(e^{(-i\hbar/2m)y^2 t}\mathcal{F}(f)(y))(x)$ makes sense as a solution of the abstract Cauchy problem (Section 10.8.1) for $\{g \in L^2_{\mathcal{C}}(R) \mid (1+y^2)\mathcal{F}(g) \in L^2_{\mathcal{C}}(R)\}$, a subspace of $L^2_{\mathcal{C}}(R)$ that contains the Schwartz space $\mathcal{S}$. Extending the operator A as previously indicated, we have

$$\mathcal{F}(A[g]) = -\frac{i\hbar y^2}{2m}\mathcal{F}(g).$$

Exercise 10.9.1. Show that $u(t)(x) = \mathcal{F}^{-1}(e^{(-i\hbar/2m)y^2 t}\mathcal{F}(f)(y))(x)$ for f a member of $\{g \in L^2_{\mathcal{C}}(R) \mid (1+y^2)\mathcal{F}(g) \in L^2_{\mathcal{C}}(R)\}$ solves the abstract Cauchy problem (Section 10.8.1).

Hint: By Plancherel's Theorem (Section 10.6.1),

$$\frac{u(t+\Delta t)(x) - u(t)(x)}{\Delta t}$$

$$= \mathcal{F}^{-1}\left(\frac{e^{(-i\hbar/2m)y^2(t+\Delta t)} - e^{(-i\hbar/2m)y^2 t}}{\Delta t} \cdot \mathcal{F}(f)(y)\right)(x)$$

$$\to \mathcal{F}^{-1}\left(\frac{-i\hbar}{2m}y^2\mathcal{F}(u(t))\right) = \mathcal{F}^{-1}(\mathcal{F}A[u(t)]) = A[u(t)],$$

since

$$\mathcal{F}(u(t)) = e^{(-i\hbar/2m)y^2 t}\mathcal{F}(f)(y) \qquad \text{and}$$

$$y^2\mathcal{F}(u(t)) = e^{(-i\hbar/2m)y^2 t}y^2\mathcal{F}(f)(y) \in L^2_{\mathcal{C}}(R),$$

because $f \in \{g \in L^2_{\mathbb{C}}(\mathbb{R}) \mid (1 + y^2)\mathcal{F}(g) \in L^2_{\mathbb{C}}(\mathbb{R})\}$. Also, for any solution v,

$$\langle A[v], v \rangle = \langle \mathcal{F}(A[v]), \mathcal{F}(v) \rangle = -\frac{i\hbar}{2m} \int_{\mathbb{R}} y^2 \, |\mathcal{F}(v)|^2 \, dy;$$

$$\mathrm{Re} \, \langle A[v], v \rangle = 0, \qquad \text{and}$$

$$\left(\|v\|^2_2 \right)' = \left(\langle v, v \rangle \right)' = 2\mathrm{Re} \, \langle v', v \rangle = 2\mathrm{Re} \, \langle A[v], v \rangle = 0.$$

The element $\|v\|^2_2$ is a constant: $\|v(t)\|^2_2 = \|v(0)\|^2_2 = 1$.

Compare this solution of the abstract Cauchy problem

$$\mathcal{F}^{-1} \left(e^{(-i\hbar/2m)y^2 t} \mathcal{F}(f)(y) \right)(x)$$

for f a member of $\{g \in L^2_{\mathbb{C}}(\mathbb{R}) \mid (1 + y^2)\mathcal{F}(g) \in L^2_{\mathbb{C}}(\mathbb{R})\}$ with

$$\left(\frac{m}{2\pi i \hbar t} \right)^{1/2} \int_{\mathbb{R}} \exp \left(\frac{im}{2\hbar t} (x - x_0)^2 \right) f(x_0) \, dx_0.$$

Mimic the argument of Example 10.9.1, starting with Part Three and f a member of $L^1_{\mathbb{C}}(\mathbb{R}) \cap L^2_{\mathbb{C}}(\mathbb{R})$. Use $L^1_{\mathbb{C}}(\mathbb{R})$ to apply the Lebesgue Dominated Convergence Theorem 6.3.3. Conclude that

$$u(t)(x) = \mathcal{F}^{-1} \left(e^{(-i\hbar/2m)y^2 t} \mathcal{F}(f)(y) \right)(x)$$

$$= \left(\frac{m}{2\pi i \hbar t} \right)^{1/2} \int_{\mathbb{R}} \exp \left(\frac{im}{2\hbar t} (x - x_0)^2 \right) f(x_0) \, dx_0$$

almost everywhere in x, for f a member of

$$L^1_{\mathbb{C}}(\mathbb{R}) \cap \{g \in L^2_{\mathbb{C}}(\mathbb{R}) \mid (1 + y^2)\mathcal{F}(g) \in L^2_{\mathbb{C}}(\mathbb{R})\} \, .$$

(The integral is an ordinary Lebesgue integral.)

Example 10.9.3. We want to remove the requirement that f be a member of $L^1_{\mathbb{C}}(\mathbb{R})$.

Note first that the integral

$$\left(\frac{m}{2\pi i \hbar t} \right)^{1/2} \int_{\mathbb{R}} \exp \left(\frac{im}{2\hbar t} (x - x_0)^2 \right) f(x_0) \, dx_0$$

$$= \left(\frac{m}{i \hbar t} \right)^{1/2} e^{(im/2\hbar t)x^2} \cdot \lim_{r \to \infty} (2\pi)^{-1/2} \int_{-r}^{r} e^{(-imx/\hbar t)x_0} \left[e^{imx_0^2/2\hbar t} f(x_0) \right] dx_0$$

$$= \left(\frac{m}{i \hbar t} \right)^{1/2} e^{imx^2/2\hbar t} \cdot \mathcal{F} \left(e^{(im(\cdot)^2/2\hbar t} f(\cdot) \right) \left(\frac{m}{\hbar t} x \right)$$

makes sense in $L^2_{\mathcal{C}}(\mathbb{R})$ ("mean"), as does $\mathcal{F}^{-1}\left(e^{(-i\hbar/2m)y^2 t}\mathcal{F}(f)(y)\right)$.
Now truncate. Given

$$f_r(x) \equiv \begin{cases} f(x) & |x| \le r, \\ 0 & \text{otherwise,} \end{cases}$$

then f_r is a member of $L^1_{\mathcal{C}}(\mathbb{R}) \cap L^2_{\mathcal{C}}(\mathbb{R})$, and

$$\begin{aligned}
u(t)(x) &= \mathcal{F}^{-1}\left(e^{(-i\hbar t/2m)y^2}\mathcal{F}(f)(y)\right)(x) \\
&= \lim_{r\to\infty} \mathcal{F}^{-1}\left(e^{(-i\hbar/2m)y^2 t}\mathcal{F}(f_r)(y)\right)(x) \qquad \text{(by 10.6.1)} \\
&= \lim_{r\to\infty} \left(\frac{m}{2\pi i\hbar t}\right)^{1/2} \int_{\mathbb{R}} \exp\left(\frac{im}{2\hbar t}(x-x_0)^2\right) f_r(x_0)\,dx_0 \\
&= \left(\frac{m}{2\pi i\hbar t}\right)^{1/2} \int_{\mathbb{R}} \exp\left(\frac{im}{2\hbar t}(x-x_0)^2\right) f(x_0)\,dx_0,
\end{aligned}$$

for f a member of $\{g \in L^2_{\mathcal{C}}(\mathbb{R}) \mid (1+y^2)\mathcal{F}(g) \in L^2_{\mathcal{C}}(\mathbb{R})\}$.

10.9.2   A Theorem

In Examples 10.9.1, 10.9.2, and 10.9.3, as well as Exercise 10.9.1, we developed the following theorem.

Theorem 10.9.1. *Given the abstract Cauchy problem* $u'(t) = A[u]$, *with* $u(0) = f$, *the solution* u *is given by*

$$\begin{aligned}
u(t)(x) &= \mathcal{F}^{-1}\left(e^{(-i\hbar/2m)y^2 t}\mathcal{F}(f)(y)\right)(x) \\
&= \left(\frac{m}{2\pi i\hbar t}\right)^{1/2} \int_{\mathbb{R}} \exp\left(\frac{im}{2\hbar t}(x-x_0)^2\right) f(x_0)\,dx_0
\end{aligned}$$

almost everywhere in x, *for the following conditions:*

1. $A[f] = \dfrac{i\hbar}{2m}\dfrac{d^2}{dx^2} f$, $D(A)$ *is the Schwartz space,* f *and* u *are members of the Schwartz space, and the "integral" is an ordinary Lebesgue integral.*

2. $A[f] = \mathcal{F}^{-1}\left(\dfrac{-i\hbar}{2m}y^2\mathcal{F}(f)(y)\right)$,
   $D(A) = \{g \in L^2_{\mathcal{C}}(\mathbb{R}) \mid (1+y^2)\mathcal{F}(g) \in L^2_{\mathcal{C}}(\mathbb{R})\}$, f *is a member of* $L^1_{\mathcal{C}}(\mathbb{R}) \cap D(A)$, *and the "integral" is an ordinary Lebesgue integral.*

3. $A[f] = \mathcal{F}^{-1}\left(\dfrac{-i\hbar}{2m}y^2\mathcal{F}(f)(y)\right)$,

$D(A) = \{g \in L^2_{\mathbb{C}}(\mathbb{R}) \mid (1 + y^2)\mathcal{F}(g) \in L^2_{\mathbb{C}}(\mathbb{R})\}$, f is a member of $D(A)$, and the integral "mean" is $\lim \int_{-r}^{r}$.

10.10   Solving Schrödinger Problem B

10.10.1   Prelude to Problem B

We are almost ready to tackle Schrödinger Problem B, for $V \neq 0$. First, observe that for f in $L^2_{\mathbb{C}}(\mathbb{R})$,

$$\mathcal{F}^{-1}\left(e^{(-i\hbar/2m)y^2t}\mathcal{F}(f)(y)\right)(x)$$
$$= \left(\frac{m}{2\pi i\hbar t}\right)^{1/2}\int_{\mathbb{R}}\exp\left(\frac{im}{2\hbar t}(x - x_0)^2\right)f(x_0)\,dx_0,$$

with "mean" integrals: $\int_{\mathbb{R}} = \lim_{r\to\infty}\int_{-r}^{r}$. Secondly, the L^2 interpretation solves the abstract Cauchy problem (Section 10.8.1) with $V = 0$ via Fourier transforms, observing the requirement that f be a member of

$$\{g \in L^2_{\mathbb{C}}(\mathbb{R}) \mid (1 + y^2)\mathcal{F}(g) \in L^2_{\mathbb{C}}(\mathbb{R})\}$$

and $\int_{\mathbb{R}}|f|^2\,dx = 1$.

More explicitly, define a family of mappings $\{F(t)\}$, for $t \geq 0$, on $L^2_{\mathbb{C}}(\mathbb{R})$ into $L^2_{\mathbb{C}}(\mathbb{R})$ by $F(0)f = f$ and

$$(F(t)f)(x) = \mathcal{F}^{-1}\left(e^{(-i\hbar/2m)y^2t}\mathcal{F}(f)(y)\right)(x)$$
$$= \left(\frac{m}{2\pi i\hbar t}\right)^{1/2}\int_{\mathbb{R}}\exp\left(\frac{im}{2\hbar t}(x - x_0)^2\right)f(x_0)\,dx_0.$$

Then

$$F\left(\frac{t}{2}\right)F\left(\frac{t}{2}\right)f$$
$$= F\left(\frac{t}{2}\right)\left(\mathcal{F}^{-1}\left(e^{(-i\hbar/2m)y^2(t/2)}\mathcal{F}(f)\right)\right)$$
$$= \mathcal{F}^{-1}\left(e^{(-i\hbar/2m)y^2(t/2)}\mathcal{F}\left(\mathcal{F}^{-1}\left(e^{(-i\hbar/2m)y^2(t/2)}\mathcal{F}(f)\right)\right)\right)$$
$$= \mathcal{F}^{-1}\left(e^{(-i\hbar/2m)y^2t}\mathcal{F}(f)\right) = F(t)f.$$

Generally, $F(t)f = \left(F(t/n)\right)^n f$, and thus

$$\left(\frac{m}{2\pi i \hbar t}\right)^{1/2} \int_{\mathrm{R}} \exp\left(\frac{im}{2\hbar t}(x - x_0)^2\right) f(x_0)\, dx_0$$

$$= \left(\frac{m}{2\pi i \hbar (t/n)}\right)^{n/2} \int_{\mathrm{R}} dx_{n-1} \cdots \int_{\mathrm{R}} dx_0$$

$$\cdot \exp\left(\frac{im}{2\hbar t/n} \sum_{1}^{n} (x_k - x_{k-1})^2\right) f(x_0),$$

which is Feynman's solution of Schrödinger Problem A — *assuming* that f is a member of $\{g \in L^2_C(\mathrm{R}) \mid (1 + y^2)\mathcal{F}(g) \in L^2_C(\mathrm{R})\}$ and $\int_{\mathrm{R}} |f|^2\, dx = 1$.

Schrödinger Problem B remains, the situation with $V \neq 0$, nonconstant coefficients.

10.10.2   Trotter's Contribution

In 1964, Edward Nelson put all the pieces together using the Trotter Product Formula (Trotter 1958, 1959). A proof may be found in Johnson and Lapidus (2000, pp. 200–201).

Theorem 10.10.1 (Trotter Product Formula). *If A, B, and A + B generate* (C_0) *contraction semigroups T, S, and U on a Hilbert space* $\mathcal{H}$, *then*

$$\lim_{n \to \infty} \left(T\left(\frac{t}{n}\right) S\left(\frac{t}{n}\right)\right)^n f = U(t)f,$$

for f a member of $\mathcal{H}$.

Trotter's product formula will provide a solution of Schrödinger Problem B, but first we have many items to define and discuss. These considerations lead to another method of solving abstract Cauchy problems, without the requirement that we have constant coefficients. For this approach, *semigroups of linear operators*, the main references are Goldstein (1985) and Johnson and Lapidus (2000).

10.10.3   Semigroups of Linear Operators

Assume a linear map $T(t)$ maps $f(x)$, the solution at time 0, to the solution at time t. We write $T(t)f = u(t)$. Thus

$$T(t + s)f = u(t + s) = T(t)u(s) = T(t)T(s)u(0) = T(t)T(s)f$$

(we hope), and $T(0)f = f$. We have a collection of linear operators in the Hilbert space $L^2_C(\mathbb{R})$ so that $T(0) = I$ and $T(t + s) = T(t)T(s)$. These ideas lead us to the notion of a *semigroup of linear operators*. Our discussion begins with the approach laid out by Martin Schechter (2001).

Suppose we want to solve the differential equation

$$u'(t) = Au(t), \qquad u(0) = u_0,$$

for $t \geq 0$, where A and u_0 are constants. Of course $u(t) = e^{tA}u_0$ is a solution for $t \geq 0$.

As before, we will abstract this problem to a new setting. Suppose we have a Banach space X, and to each real variable t we assign a function $u(t)$ in X. We can think of A as a mapping of X into X. That is, for u in X, Au belongs to X. We interpret $u'(t) = Au(t)$ as

$$\left\| \frac{u(t + h) - u(t)}{h} - Au(t) \right\| \to 0 \text{ as } h \to 0.$$

In this new setting, can we make sense out of e^{tA} when A is an operator?

Naturally, we try

$$e^{tA} = I + \frac{1}{1!}tA + \frac{1}{2!}t^2A^2 + \cdots,$$

and with $u(t)$ a member of X, we have

$$u(t) = e^{tA}u(0) = \left(I + \frac{1}{1!}tA + \frac{1}{2!}t^2A^2 + \cdots \right)u(0).$$

Thus

$$\frac{u(t + h) - u(t)}{h} = \frac{e^{(t+h)A} - e^{tA}}{h}u(0) = \left(\frac{e^{hA} - I}{h} \right)e^{tA}u(0),$$

and

$$u'(t) = \lim_{h \to 0} \left(\frac{e^{hA} - I}{h} \right)e^{tA}u(0) = Ae^{tA}u(0) = Au(t).$$

More questions suggest themselves:

- Can we justify these manipulations?

- Does $\left(I + \frac{1}{1!}tA + \frac{1}{2!}t^2A^2 + \cdots \right)u(0)$ converge to an element of X?

- Does $e^{(t+h)A} = e^{tA} \cdot e^{hA} = e^{hA} \cdot e^{tA}$ make any sense?

Suppose A is a bounded linear operator in X and thus continuous. Then for $N > M$ and $t > 0$,

$$\left\| \sum_M^N \frac{1}{k!} t^k A^k \right\| \leq \sum_M^N \frac{1}{k!} t^k \|A\|^k \to 0 \quad \text{as } M, N \to \infty.$$

Defining e^{tA} by $I + \frac{1}{1!} tA + \frac{1}{2!} t^2 A^2 + \cdots$ makes sense, and $\|e^{tA}\| \leq e^{t\|A\|}$ for $t > 0$.

Next,

$$\frac{e^{hA} - I}{h} = \sum_1^\infty \frac{1}{k!} h^{k-1} A^k,$$

and

$$\left\| \frac{e^{hA} - I}{h} - A \right\| \leq \sum_2 \frac{1}{k!} h^{k-1} \|A\|^{k-1} = \frac{e^{h\|A\|} - I}{h} - \|A\| \to 0 \quad \text{as } h \to 0.$$

Thus $u'(t) = Ae^{tA} u(0)$.

But we still need to discuss whether $e^{(t+h)A} = e^{tA} \cdot e^{hA} = e^{hA} \cdot e^{tA}$. Is it true in some sense that

$$\sum \frac{1}{k!} (t+h)^k A^k = \left(\sum \frac{1}{m!} t^m A^m \right) \left(\sum \frac{1}{n!} h^n A^n \right)$$

$$= \left(\sum \frac{1}{h!} h^n A^n \right) \left(\sum \frac{1}{m!} t^m A^m \right)?$$

For real numbers,

$$(x+y)^k = \sum_{m=0}^k \frac{k!}{m!(k-m)!} x^m y^{k-m} \qquad \text{and}$$

$$\sum_0 \frac{(x+y)^k}{k!} = \left(\sum_0 \frac{x^n}{n!} \right) \left(\sum_0 \frac{y^m}{m!} \right).$$

How does this fare in our setting? We have

$$\sum_{k=0}^N \frac{1}{k!} (t+h)^k A^k = \sum_{k=0}^N \frac{1}{k!} \left(\sum_{m=0}^k \frac{k!}{m!(k-m)!} t^m h^{k-m} \right) A^k$$

$$= \sum_{k=0}^N \sum_{m=0}^k \frac{1}{m!(k-m)!} t^m A^m h^{k-m} A^{k-m}$$

$$= \sum_{m+n \leq N} \frac{1}{m!n!} t^m A^m h^n A^n.$$

Thus

$$\left\| \sum_{m=0}^{N} \frac{t^m}{m!} A^m \cdot \sum_{n=0}^{N} \frac{h^n}{n!} A^n - \sum_{k=0}^{N} \frac{1}{k!} (t+h)^k A^k \right\|$$

$$= \left\| \sum_{m=0}^{N} \frac{1}{m!} t^m A^m \cdot \sum_{n=0}^{N} \frac{1}{n!} h^n A^n - \sum_{m+n \le N} \frac{1}{m!n!} t^m A^m h^n A^n \right\|$$

$$= \left\| \sum_{\substack{m,n \le N \\ n+m > N}} \frac{1}{m!n!} t^m A^m h^n A^n \right\|$$

$$\le \sum_{\substack{n,m \le N \\ n+m > N}} \frac{1}{n!m!} t^m \|A\|^m h^n \|A\|^n$$

$$= \sum_{m=0}^{N} \frac{1}{m!} t^m \|A\|^m \cdot \sum_{n=0}^{N} \frac{1}{n!} h^n \|A\|^n - \sum_{k=0}^{N} \frac{1}{k!} (t+h)^k \|A\|^k$$

$$\to e^{t\|A\|} e^{h\|A\|} - e^{(t+h)\|A\|} = 0.$$

This is very encouraging.

Unfortunately, in our particular application A is not bounded; A is a differential operator. However, loosely speaking (A as a limit of bounded linear operators) linearity of A on a dense subspace of X is good enough.

10.10.4   Semigroup Terminology and a Theorem

Before we proceed, it is necessary to make some definitions concerning a family $T = \{T(t) : 0 \le t < \infty\}$ of everywhere defined bounded linear operators from a Banach space X to itself.

Definition 10.10.1 (Semigroup). The family T is called a (C_0) semigroup iff

1. $T(t+s) = T(t)T(s)f$, with $t, s \ge 0$, for all f in X.

2. $T(0)f = f$.

3. The map from $[0, \infty)$ to X, $t \to T(t)f$ is continuous in the norm of X for each f in X.

The third criterion is called the *strong operator continuity* of the semigroup: $|t - s|$ small implies $\|T(t)f - T(s)f\|_X = \|(T(t) - T(s))f\|_X$ small.

Definition 10.10.2 (Contraction Semigroup). The family T is called a (C_0) contraction semigroup if, in addition, a fourth criterion holds:

4. $\|T(t)f\| \leq \|f\|$ for all $t \geq 0$ and all f in X.

Definition 10.10.3 (Generator of a Semigroup). The generator A of a (C_0) semigroup T is defined by

$$A[f] \equiv \lim_{t \to 0^+} \frac{1}{t}[T(t)f - f] = \lim_{t \to 0^+} \frac{(T(t) - I)f}{t} \equiv T'(0)f,$$

where the domain of A, $D(A)$, consists of those elements of X for which this limit exists.

A (C_0) semigroup with generator A solves the abstract Cauchy problem, in the sense we defined it in Section 10.8.1. We can make sense out of $u(t) = e^{tA}u_0$.

Theorem 10.10.2. *Let $\{T(t)\}$ be a (C_0) semigroup on the Banach space X with generator A, and let f be a member of the domain of A. That is,*

$$A[f] \equiv \lim_{t \to 0^+} \frac{T(t)f - f}{t} = \frac{d}{dt}T(t)f \Big|_{t=0} = T'(0)f$$

exists. Then the domain of A is a dense subspace of X, and $u(t) \equiv T(t)f$ is the unique continuously differentiable solution on $[0, \infty)$ of the abstract Cauchy problem for $t \geq 0$,

$$u'(t) = A[u(t)], \qquad u(0) = f.$$

Proof. Details can be found in Goldstein (1985) and in Johnson and Lapidus (2000). We argue only existence, in six steps.

Step 1. $u(0) \equiv T(0)f = f$.

Step 2. The domain of A is a linear subspace of X, and A is a linear operator on this subspace of X. (By definition we have linearity of T on the Banach space X.)

Step 3. The domain of A is a dense subspace of X. Let f be a member of X. Form the Riemann-type integral $1/t \int_0^t T(s)f \, ds$, for $t > 0$. These Riemann-type sums converge in the norm of X. This integral makes sense because $s \to T(s)f$ is a continuous map form $[0, \infty)$ to X by assumption.

Let h be greater than 0. Then

$$\frac{T(h)\left(\frac{1}{t}\int_0^t T(s)f\,ds\right) - \left(\frac{1}{t}\int_0^t T(s)f\,ds\right)}{h}$$

$$= \frac{1}{t}\left\{\frac{1}{h}\int_0^t T(h+s)f\,ds - \frac{1}{h}\int_0^t T(s)f\,ds\right\}$$

$$= \frac{1}{t}\left\{\frac{1}{h}\int_t^{t+h} T(u)f\,du - \frac{1}{h}\int_0^h T(s)f\,ds\right\}$$

$$\rightarrow \frac{T(t)f - f}{t} \quad \text{as } h \rightarrow 0^+.$$

Thus $1/t\int_0^t T(s)f\,ds$ belongs to the domain of A, and

$$A\left[\frac{1}{t}\int_0^t T(s)f\,ds\right] = \frac{T(t)f - f}{t}.$$

On the other hand,

$$\frac{1}{t}\int_0^t T(s)f\,ds \rightarrow T(0)f = f \quad \text{as } t \rightarrow 0^+.$$

Given an arbitrary element f of X, we have an element in the domain of A, $1/t\int_0^t T(s)f\,ds$ arbitrarily close for t sufficiently small. The domain of A is a dense subspace of X.

Step 4. The operator $T(t)$ maps the domain of A into A, and $T(t)A[f] = A[T(t)f]$ for f in the domain of A. Suppose f belongs to the domain of A. We show that $T(t)f$ belongs to the domain of A:

$$A[T(t)f] = \lim_{h\rightarrow 0^+} \frac{T(h)T(t)f - T(t)f}{h} = \lim_{h\rightarrow 0^+} T(t)\frac{(T(h)-I)f}{h}$$

$$= T(t)A[f].$$

We know that

$$\lim_{h\rightarrow 0^+} \frac{(T(h)-I)f}{h} = A[f]$$

because by assumption f belongs to the domain of A. Therefore the limit exists and $A[T(t)f] = T(t)A[f]$. The family T commutes with its generator A.

Step 5. For any interval $[a, b]$ of $[0, \infty)$, we have a constant M such that $\|T(t)\| \leq M$ for all t in the interval $[a, b]$. Because T is a (C_0) semigroup, $s \to T(s)f$ is a continuous function from $[0, \infty)$ to X for a fixed $f \in X$. So $s \to \|T(s)f\|$ is a continuous function from $[0, \infty)$ to R for a fixed $f \in X$.

We have a constant M_f, so that $\|T(s)f\| \leq M_f$ with $a \leq s \leq b$, for each $f \in X$. By the Uniform Boundedness Principle, we have a constant M so that $\|T(s)\| \leq M$ for $a \leq s \leq b$.

Step 6. We claim $u(t) \equiv T(t)f$ is a solution of $u'(t) = A[u(t)]$ for any f in the domain of A. That is, for any member f of the dense domain of A, we have a solution, $T(t)f$, of the differential equation $u' = A[u]$. The argument is as follows.

Let f belong to the domain of A, and suppose $t > 0$. Then

$$\lim_{h \to 0^+} \frac{T(t+h)f - T(t)f}{h} = \lim_{h \to 0^+} \frac{T(h) - I}{h} T(t)f = A[T(t)f],$$

because $T(t)f$ belongs to the domain of A by Step 4. Furthermore,

$$\lim_{h \to 0^+} \frac{T(t)f - T(t-h)f}{h} = \lim_{h \to 0^+} T(t-h)\frac{(T(h) - I)f}{h} = T(t)A[f],$$

because f belongs to the domain of A by assumption.

For the last equality, observe that

$$\left\| T(t-h)\frac{(T(h) - I)f}{h} - T(t)A[f] \right\|_X$$

$$\leq \left\| T(t-h)\left\{ \frac{(T(h) - I)f}{h} - A[f] \right\} \right\|_X + \|(T(t-h) - T(t))A[f]\|_X$$

$$\leq M \left\| \frac{(T(h) - I)f}{h} - A[f] \right\|_X + \|(T(t-h) - T(t))A[f]\|_X.$$

As $t \to T(t)A[f]$ is a continuous function from $[0, \infty)$ to X, this allows us to estimate $\|(T(t-h) - T(t))A[f]\|_X$. By definition,

$$\lim_{h \to 0^+} \frac{(T(h) - I)f}{h} = A[f],$$

because f belongs to the domain of A. Since by Step 4 $T(t)A[f] = A[T(t)f]$, we may conclude that $(T(t)f)' = A[T(t)f]$.

Thus $u(t) \equiv T(t)f$ is a solution of the differentiatial equation $u'(t) = A[u(t)]$. In fact, because $(T(t)f)' = T(t)A[f]$, the function $u(t)$ is continuously differentiable for $t \geq 0$. $\square$

We have shown that $u(t) \equiv T(t)f$ is a solution of the abstract Cauchy problem according to our definition (Section 10.8.1).

10.10.5 Some Notes on Our Solution

Note 1.

$$T(t)f - f = \int_0^t \frac{d}{ds}(T(s)f)ds = \int_0^t A[T(s)f]dx$$
$$= \int_0^t T(s)A[f]ds$$

for f in the domain of A.

Note 2. Suppose the sequence $\{f_n\}$ of members of the domain of A converges to f; that is, suppose $\|f_n - f\|_X \to 0$. Likewise, say the sequence $\{Af_n\}$ converges to g, so $\|Af_n - g\|_X \to 0$. We claim that f belongs to the domain of A and that $Af = g$.

By Step 5 we have $\|T(t)\| \leq M$ for $0 \leq t \leq T$. To show that f belongs to the domain of A, we form the quotient

$$\frac{T(t)f - f}{t} = \lim \frac{T(t)f_n - f_n}{t} = \lim \frac{1}{t}\int_0^t T(s)A[f_n]ds$$
$$= \frac{1}{t}\int_0^t T(s)\lim A[f_n]ds = \frac{1}{t}\int_0^t T(s)g\,ds.$$

We can do so because

$$\left\| \frac{1}{t}\int_0^t [T(s)A[f_n] - T(s)g]ds \right\|_X = \left\| \frac{1}{t}\int_0^t T(s)[A[f_n] - g]ds \right\|_X$$
$$\leq M\frac{1}{t}\int_0^t \|A[f_n] - g\|_X\,ds$$
$$\leq M\|A[f_n] - g\|_X.$$

Thus

$$\lim_{t \to 0^+} \frac{T(t)f - f}{t} = \lim_{t \to 0^+} \frac{1}{t}\int_0^t T(s)g\,ds = T(0)g = g.$$

Hence f belongs to the domain of A, and $Af = g$. We refer to A as a *closed operator* on the Banach space X.

Note 3. The solution $u(t) \equiv T(t)f$ is the unique solution to the abstract Cauchy problem $u'(t) = A[u(t)]$, for $t \geq 0$ and $u(0) = f$ for f in the domain of A.

10.10.6   Applying the Theorem

To aid our understanding of Theorem 10.10.2, we will mimic the argument for a specific T. For f a member of $L^2_{\mathbb{C}}(\mathcal{R})$, define for $t > 0$,

$$\left(T(t)f\right)(x) \equiv \mathcal{F}^{-1}\left(e^{(-i\hbar/2m)y^2 t}\mathcal{F}(f)\right)(x), \qquad T(0)f = f.$$

We proceed as before.

Step 1. By Plancherel's Theorem (Section 10.6.1), T is a linear operator from $L^2_{\mathbb{C}}(\mathcal{R})$ into $L^2_{\mathbb{C}}(\mathcal{R})$.

Step 2. Again by Plancherel's Theorem,

$$\begin{aligned}
\|T(t)f\|^2 &= \left\langle \mathcal{F}^{-1}\left(e^{(-i\hbar/2m)y^2 t}\mathcal{F}(f)\right), \mathcal{F}^{-1}\left(e^{(-i\hbar/2m)y^2 t}\mathcal{F}(f)\right)\right\rangle \\
&= \langle e^{(-i\hbar/2m)y^2 t}\mathcal{F}(f), e^{(-i\hbar/2m)y^2 t}\mathcal{F}(f)\rangle \\
&= \langle \mathcal{F}(f), \mathcal{F}(f)\rangle = \|f\|^2.
\end{aligned}$$

We have that T is bounded: $\|T(t)\| = 1$ for all $t \geq 0$. We have a family $T = \{T(t): 0 \leq t < \infty\}$ of everywhere defined bounded linear operators in $L^2_{\mathbb{C}}(\mathcal{R})$.

Step 3. We calculate

$$\begin{aligned}
\left(T(t)T(s)f\right) &= T(t)\left(\mathcal{F}^{-1}\left(e^{(-i\hbar/2m)y^2 s}\mathcal{F}(f)\right)\right) \\
&= \mathcal{F}^{-1}\left(e^{(-i\hbar/2m)y^2 t}\mathcal{F}\left(\mathcal{F}^{-1}\left(e^{(-i\hbar/2m)y^2 s}\mathcal{F}(f)\right)\right)\right) \\
&= \mathcal{F}^{-1}\left(e^{(-i\hbar/2m)y^2(t+s)}\mathcal{F}(f)\right) = T(t+s)f.
\end{aligned}$$

Step 4. Plancherel's Theorem gives us

$$\begin{aligned}
T(t+\Delta t)f &= T(t)T(\Delta t)f = \mathcal{F}^{-1}\left(e^{(-i\hbar/2m)y^2(t+\Delta t)}\mathcal{F}(f)\right) \\
&\xrightarrow[\Delta t \to 0]{} \mathcal{F}^{-1}\left(e^{(-i\hbar/2m)y^2 t}\mathcal{F}(f)\right) = T(t)f.
\end{aligned}$$

We have a (C_0) contraction semigroup T. The generator of T is given by

$$A[f] \equiv \lim_{t \to 0^+} \frac{T(t)f - f}{t},$$

with the domain of A being those elements of $L^2_C(\mathcal{R})$ for which this limit makes sense. However (again invoking Plancherel's Theorem),

$$\lim_{t \to 0^+} \mathcal{F}^{-1}\left(\left(\frac{e^{(-i\hbar/2m)y^2 t} - 1}{t}\right)\mathcal{F}(f)\right)$$

$$= \lim_{t \to 0^+} \mathcal{F}^{-1}\left(e^{(-i\hbar/2m)y^2(t/2)}\left[-2i\left(\frac{\sin \frac{\hbar}{2m}y^2\frac{t}{2}}{\frac{\hbar}{2m}y^2\frac{t}{2}}\right) \cdot \frac{\hbar}{2m}\frac{y^2}{2}\right]\mathcal{F}(f)\right)$$

$$= \mathcal{F}^{-1}\left(\frac{-i\hbar}{2m}y^2\mathcal{F}(f)\right)$$

for $(1 + y^2)\mathcal{F}(f)$ in $L^2_C(R)$. Yes,

$$A[f] = \mathcal{F}^{-1}\left(\frac{-i\hbar}{2m}y^2\mathcal{F}(f)\right)$$

for $D(A) = \{g \in L^2_C(R) \mid (1 + y^2)\mathcal{F}(g) \in L^2_C(R)\}$.

Step 5. The domain of A is a (dense) subspace of $L^2_C(R)$. It contains the Schwartz space.

Step 6. We noted that

$$A[T(t)f] = \mathcal{F}^{-1}\left(\frac{-i\hbar}{2m}y^2\mathcal{F}(T(t)f)\right)$$

$$= \mathcal{F}^{-1}\left(\frac{-i\hbar}{2m}y^2\mathcal{F}\left(\mathcal{F}^{-1}(e^{(-i\hbar/2m)y^2 t}\mathcal{F}(f))\right)\right)$$

$$= \mathcal{F}^{-1}\left(\frac{-i\hbar}{2m}y^2 e^{(-i\hbar/2m)y^2 t}\mathcal{F}(f)\right)$$

$$= \mathcal{F}^{-1}\left(e^{(-i\hbar/2m)y^2 t} \cdot \frac{-i\hbar}{2m}y^2\mathcal{F}(f)\right) = T(t)A[f],$$

and so on. Thus for f a member of $\{g \in L^2_C(R) \mid (1 + y^2)\mathcal{F}(g) \in L^2_C(R)\}$,

$$u(t)(x) \equiv \mathcal{F}^{-1}(e^{(-i\hbar/2m)y^2 t}\mathcal{F}(f)y)(x)$$

$$= \left(\frac{m}{2\pi i\hbar t}\right)^{1/2}\int_R \exp\left(\frac{im}{2\hbar t}(x - x_0)^2\right)f(x_0)\,dx_0$$

(almost everywhere in x, "mean" integral) solves the abstract Cauchy problem $\dfrac{du}{dt} = A[u]$ with $u(0) = f$.

Step 7. For f a member of $L_C^2(\mathbb{R})$, for $t > 0$,

$$(T(t)f)(x) \equiv \mathcal{F}^{-1}\left(e^{(-i\hbar/2m)y^2 t} \mathcal{F}(f)(y)\right)(x)$$

$$= \left(\frac{m}{2\pi i \hbar t}\right)^{1/2} \int_{\mathbb{R}} e^{(im/2\hbar t)(x-x_0)^2} f(x_0)\, dx_0,$$

and $T(0)f = f$. Thus the family T is a (C_0) contraction semigroup on $L_C^2(\mathbb{R})$, and the generator of T is A.

10.10.7   Problem B and the Trotter Product

Because Schrödinger's Equation

$$\frac{\partial \psi}{\partial t} = \frac{i\hbar}{2m}\frac{\partial^2 \psi}{\partial x^2} - \frac{i}{\hbar}V\psi$$

and the operator A dealt with

$$\frac{i\hbar}{2m}\frac{\partial^2}{\partial x^2},$$

we are led to the operator of multiplication, $B = (-i/\hbar)V$. Define

$$(S(t)f)(x) \underset{D}{\equiv} e^{(-i/\hbar)V(x)t} \cdot f(x)$$

for $t > 0$, and $S(0)f = f$ for f a member of $L_C^2(\mathbb{R})$, with V a real-valued Lebesgue measurable function.

1. S is a linear operator from $L_C^2(\mathbb{R})$ into $L_C^2(\mathbb{R})$.

2. $\|S(t)f\|^2 = \|f\|^2$, S is bounded, and $\|S(t)\| = 1$.

3. $S(t+s)f = S(t)S(s)f$.

4. $S(t+\Delta t)f = S(t)S(\Delta t)f = e^{(-i\hbar/2m)V(t+\Delta t)} \cdot f \to_{\Delta t \to 0} S(t)f$.

We have a (C_0) contraction semigroup S. The generator B of S is given by

$$B[f] \underset{0}{\equiv} \lim_{t \to 0^+} \frac{S(t)f - f}{t} = \frac{-i}{\hbar}V(x)f.$$

The operator B is linear, and its domain, a subspace of $L^2_{\mathcal{C}}(R)$, consists of the elements of $L^2_{\mathcal{C}}(R)$ that, upon multiplication by V, remain a member of $L^2_{\mathcal{C}}(R)$. The domain of B contains those infinitely differentiable functions of compact support, a dense subspace of $L^2_{\mathcal{C}}(R)$.

We have two contraction semigroups on $L^2_{\mathcal{C}}(R)$. For f a member of $L^2_{\mathcal{C}}(R)$, $T(t)f = \mathcal{F}^{-1}\left(e^{(-i\hbar/2m)y^2 t}\mathcal{F}\right)$ with generator

$$A[f] = \mathcal{F}^{-1}\left(\frac{-i\hbar}{2m}y^2\mathcal{F}(f)\right),$$

and $D(A) = \{g \in L^2_{\mathcal{C}}(R) \mid (1 + y^2)\mathcal{F}(g) \in L^2_{\mathcal{C}}(R)\}$. We have $S(t)f = e^{(-i/\hbar)Vt} \cdot f$, for V a real-valued Lebesgue measurable function with generator

$$B[f] = \frac{-i}{\hbar}V \cdot f,$$

and $D(B) = \{g \in L^2_{\mathcal{C}}(R) \mid V \cdot g \in L^2_{\mathcal{C}}(R)\}$.

Form $T(t/n)S(t/n)f$:

$$\left(T\left(\frac{t}{n}\right)S\left(\frac{t}{n}\right)f\right)(x_1)$$

$$= T\left(\frac{t}{n}\right)\left(e^{(-i/\hbar)V(t/n)}f\right)(x_1)$$

$$= \left(\frac{m}{2\pi i\hbar(t/n)}\right)^{1/2}\int_R dx_0 \exp\left(\frac{im}{2\hbar(t/n)}(x_1 - x_0)^2\right)$$

$$\cdot \exp\left(\frac{-i}{\hbar}V(x_0)\frac{t}{n}\right)f(x_0)$$

$$= \left(\frac{m}{2\pi i\hbar(t/n)}\right)^{1/2}\int_R dx_0$$

$$\cdot \exp\left(\frac{i}{\hbar}\left(\frac{m}{2t/n}(x_1 - x_0)^2 - V(x_0)\frac{t}{n}\right)\right)f(x_0)$$

almost everywhere in x_1.

In general, for $f \in L^2_{\mathcal{C}}(R)$,

$$\left(\left[T\left(\frac{t}{n}\right)S\left(\frac{t}{n}\right)\right]^n f\right)(x_n)$$

$$= \left(\frac{m}{2\pi i\hbar(t/n)}\right)^{n/2}\int_R dx_{n-1}\cdots\int_R dx_0$$

$$\cdot \exp\left(\frac{im}{2\hbar(t/n)}\sum_1^n(x_k - x_{k-1})^2 - \frac{i}{\hbar}\frac{t}{n}\sum_1^n V(x_{k-1})\right)f(x_0)$$

defines a function in $L^2_{\mathcal{C}}(R)$, with V a real-valued Lebesgue measurable function.

Consider the operator $A + B$ with domain $D(A) \cap D(B)$.

- Is this operator the generator of a (C_0) contraction semigroup?

- Is $D(A) \cap D(B)$ a dense subspace of $L^2_{\mathcal{C}}(R)$?

T. Kato (1951) showed that, in particular, if the real-valued Lebesgue measurable function V is in $L^2_R(R)$, then $D(B) \supset D(A)$, and the operator $A + B$ is the generator of a (C_0) contraction semigroup.

Apply the Trotter Product Formula (Theorem 10.10.1):

$$\left(\left[T\left(\frac{t}{n}\right) S\left(\frac{t}{n}\right) \right]^n f \right)(x)$$

$$= \left(\frac{m}{2\pi i \hbar(t/n)} \right)^{n/2} \int_R dx_{n-1} \cdots \int_R dx_0$$

$$\cdot \exp\left(\frac{im}{2\hbar(t/n)} \sum_1^n (x_k - x_{k-1})^2 - \frac{i}{\hbar} \frac{t}{n} \sum_1^n V(x_{k-1}) \right) f(x_0)$$

is a member of $L^2_{\mathcal{C}}(R)$, say $\psi_n(t, x)$, almost everywhere in $x = x_n$, with "mean" integrals. Thus we have a function in $L^2_{\mathcal{C}}(R)$, $\psi(t, x)$, and $L^2_{\mathcal{C}}(R)$ convergence. That is,

$$\|\psi_n(t, x) - \psi(t, x)\|^2 \to 0 \quad \text{as } n \to \infty.$$

So,

$$\lim_{n \to \infty} \left(\frac{m}{2\pi i \hbar(t/n)} \right)^{n/2} \int_R dx_{n-1} \cdots \int_R dx_0$$

$$\exp\left(\frac{in}{2\hbar(t/n)} \sum_1^n (x_k - x_{k-1})^2 - \frac{i}{\hbar} \frac{t}{n} \sum_1^n V(x_{k-1}) \right) f(x_0)$$

solves Schrödinger Problem B under the assumption that f is a member of

$$\left\{ g \in L^2_{\mathcal{C}}(R) \mid (1 + y^2) \mathcal{F}(g) \in L^2_{\mathcal{C}}(R) \right\},$$

V is a member of $L^2_R(R)$, and $\int_R |f|^2 \, dx = 1$. Compare with Feynman's solution to Schrödinger Problem B.

This concludes our treatment of the Feynman integral.

10.11 References

1. Fattorini, Hector. *The Cauchy Problem*. Reading, Mass.: Addison-Wesley, 1983.

2. Feynman, Richard P., and Albert R. Hibbs. *Quantum Mechanics and Path Integrals*. New York: McGraw-Hill, 1965.

3. Goldstein, Jerome. *Semigroups of Linear Operators and Applications*. Oxford University Press, 1985.

4. Johnson, Gerald, and Michel Lapidus. *The Feynman Integral and Feynman's Operational Calculus*. Oxford University Press, 2000.

5. Kato, T. Fundamental properties of Hamiltonian operators of Schrödinger types. *Transactions of the American Mathematical Society* 70 (1951) 195–211.

6. Nelson, Edward. Feynman integral and Schrödinger equation. *Journal of Mathematical Physics* 5 (1964) 332–343.

7. Schechter, Martin. *Principles of Functional Analysis*. Graduate Studies in Mathematics, Vol. 36. Providence, R.I.: American Mathematical Society, 2001.

8. Trotter, Hale F. Approximation of semigroups of operators. *Pacific Journal of Mathematics* 8 (1958) 887–920.

9. ———. On the product of semigroups of operators. *Proceedings of the American Mathematical Society* 10 (1959) 545–51.

10. Weidmann, Joachim. *Linear Operators in Hilbert Spaces*. New York: Springer-Verlag, 1980.

No matter how far we go into the future there will always be new things happening, new information coming in, new worlds to explore, a constantly expanding domain of life, consciousness, and memory.

— Freeman Dyson

Index

About the Author

Frank Burk was raised in Liberty, Missouri, and graduated from Liberty High School in 1960, a Merit Scholar. In 1963 he graduated from the University of Missouri with a BA in mathematics. He earned his PhD in mathematics from the University of California, Riverside in 1969.

Frank joined the faculty at California State University, Chico as a member of the mathematics department in 1968. He taught 37 years before retiring in 2004. He also taught for Butte College. Frank was the author of two books, this one and *Lebesgue Measure and Integration* (Wiley, 1998). He was a long-time member of the Mathematical Association of America.

Frank was a man of many interests. He was a runner his entire adult life and was one of the founders of the Chico Running Club. Frank earned the rank of Eagle Scout in his youth and continued in scouting as scoutmaster of Troop 230 for 18 years. Frank and his wife were foster parents. He was a proud member of the Greater Chico Kiwanis Club. He was a backpacker and hiked the entire John Muir Trail in 2001. Frank enjoyed reading, gardening, and playing bridge.

After a losing battle with cancer, Frank passed away Saturday, March 17, 2007, at his home in Chico.